i

A LETTER FROM JACK DANGERMOND

Each year we showcase some of the work of our users in a formal publication known as the Map Book. This year the 2020 *Esri Map Book* (Volume 36), illustrates the diversity and scope of their applications. I want to both acknowledge and congratulate every contributor. Their work shows how the application of GIS is creating positive outcomes in organizations and across society.

While geographic information systems (GIS) continue to grow on many fronts, in the past year the power and importance of GIS was dramatically demonstrated in its application to the coronavirus disease 2019 (COVID-19) pandemic. Thousands of organizations built GIS-based systems with dashboards and mapping/reporting capabilities that helped support medical operations and kept the public informed.

This work fundamentally changed how people saw the world, improving spatial literacy.

GIS professionals around the world continue helping to improve efficiency, communication, and better decision-makers. Their work is also creating more collaboration and increasingly engaging senior policy and decision-makers, helping them better understand and act using geographic science-based information.

This work inspires and motivates me and everyone at Esri. We can see how the advances we are making in GIS are empowering our users and allowing them to better advance the power of geography in their work. It is our hope that this, in turn, will help create a better and more sustainable world.

Warm regards,

Jack Dangermond

CONTENTS

HEDONIC PRICE MODEL IN REAL ESTATE VALUATION IN A POST-COMMUNIST CITY

Universitatea Alexandru Ioan Cuza, Iaşi, Romania
By Oliver-Valentin Dinter

Price per Square Meter of the Apartments in Iaşi Municipality

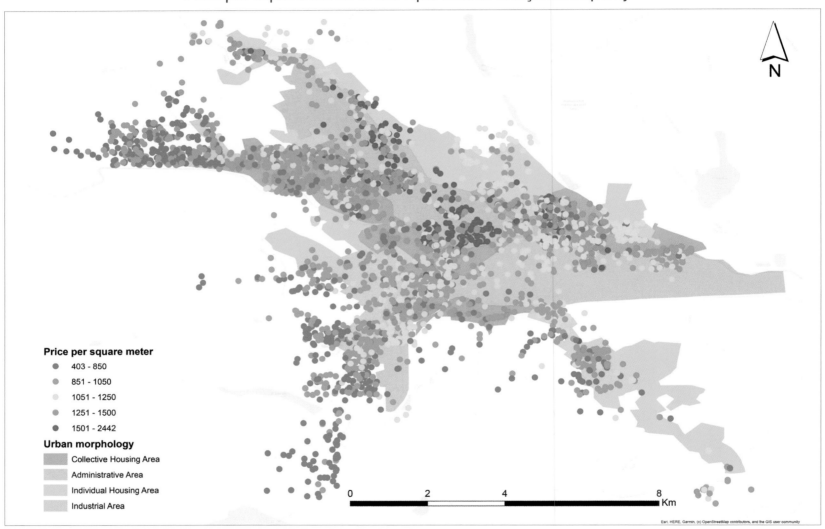

Courtesy of of Alexandru Ioan Cuza, University of Iaşi

Often described as the Romanian Capital of Culture and officially the Historic Capital of Romania, the city of Iaşi (500,000 inhabitants) also boasts the title of the Emerging City of the Year (2018) and the second-best city in Europe regarding cost-effectiveness (2020/2021). Still, the present landscape was heavily shaped by the communist regime and post-socialist transition. Recently, massive investments in urban renewal near the historic city center led to an explosion of prices per square meter in the heart of the city. To understand inequalities across the city, a regression was computed using explanatory housing attributes such as surface area, number of rooms, floor level, year of construction, and spatial attributes such as accessibility to eight categories of nearby facilities (schools, bus stops, office buildings, and so on). The above-mentioned attributes are essential in the hedonic price model, focusing on the factors that increase the satisfaction of the potential buyer. Residual values were clustered to reveal areas of overvalued and undervalued apartments. As a result, the second map shows that

The Impact of Unquantifiable Housing Attibutes on Real Estate Valuation

Price clusters

- Not Significant
- Overvalued apartment
- Overvalued apartment in an undervalued area
- Undervalued apartment in an overvalued area
- Undervalued apartment

Urban morphology

- Collective Housing Area
- Administrative Area
- Individual Housing Area
- Industrial Area

| 0 | 2 | 4 | 8 |
Km

Esri, HERE, Garmin, (c) OpenStreetMap contributors, and the GIS user community

the administrative area (mostly in the city center) is overvalued because of the incidence of unquantifiable attributes that couldn't be included in the regression but record a great influence on the price per square meter (prestige of the area—the prevalence of cultural monuments, top high schools, pedestrian facilities, and so on). On the other side, undervalued apartments are located closer to the suburbs, and the low level of infrastructure, lack of facilities, and heavy traffic are responsible for the low price.

CONTACT
Oliver-Valentin Dinter
oliverdinter7@gmail.com

SOFTWARE
ArcGIS® Desktop 10.2

DATA SOURCES
imobiliare.ro, openstreetmap.org

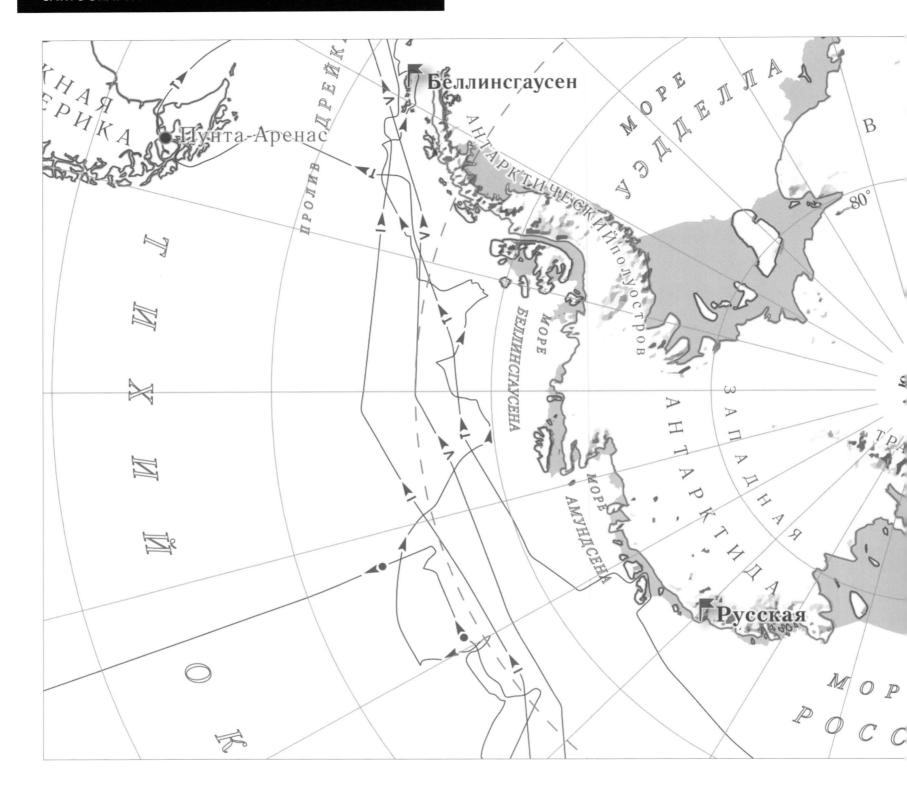

RUSSIAN EXPLORATION OF ANTARCTICA

Technical University Munich, Munich, Germany
By Nikita Slavin

An overview map of historical Russian expeditions to Antarctica. This map, using shades of only one color, was inspired by the MonoCarto mapping design competition organized by Daniel P. Huffman. The style of the map reflects Soviet maps of the Antarctic. This map shows the main bases and routes of Russian and Soviet expeditions across the Antarctic continent. The challenge was to make all the different tracks distinguishable using arrows and dashes instead of different color symbology. The routes are complemented with silhouettes of the ships.

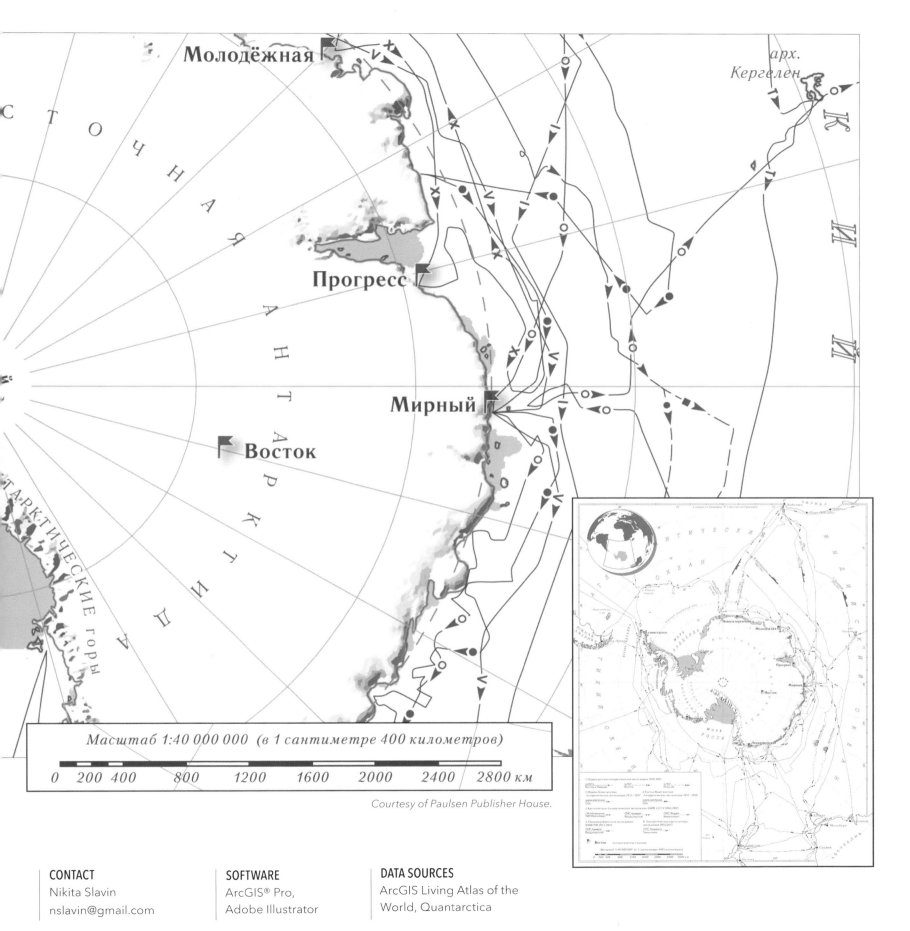

Масштаб 1:40 000 000 (в 1 сантиметре 400 километров)

0   200 400   800   1200   1600   2000   2400   2800 км

Молодёжная

арх.
Кергелен

Прогресс

Мирный

Восток

Courtesy of Paulsen Publisher House.

CONTACT
Nikita Slavin
nslavin@gmail.com

SOFTWARE
ArcGIS® Pro,
Adobe Illustrator

DATA SOURCES
ArcGIS Living Atlas of the
World, Quantarctica

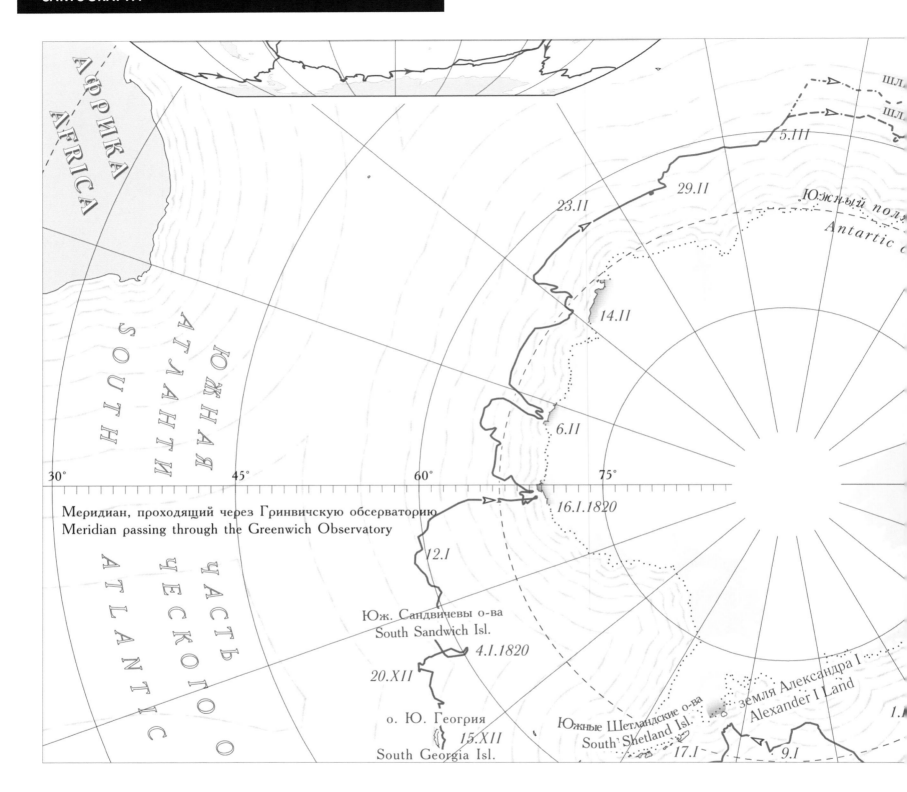

FIRST RUSSIAN ANTARCTIC EXPEDITION MAP

Technical University Munich,
Munich, Germany
By Nikita Slavin

A bilingual map in Russian and English of the expedition route of Thaddeus Bellingshausen and Mikhail Lazarev, made for the anniversary reprint of *The Voyage of Captain Bellingshausen to the Antarctic Seas*.

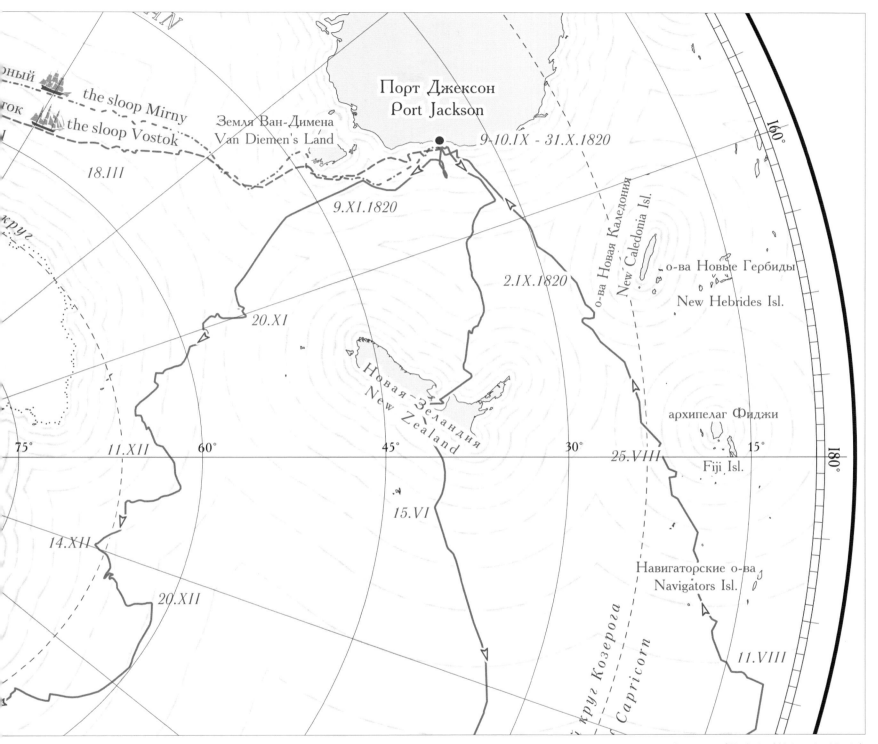

онный

the sloop Mirny

ток

the sloop Vostok

18.III

Земля Ван-Димена
Van Diemen's Land

Порт Джексон
Port Jackson

9-10.IX - 31.X.1820

9.XI.1820

2.IX.1820

о-ва Новая Каледония
New Caledonia Isl.

о-ва Новые Гербиды
New Hebrides Isl.

круг

20.XI

Новая-Зеландия
New Zealand

архипелаг Фиджи

75°

11.XII

60°

45°

30°

25.VIII

15°

Fiji Isl.

15.VI

14.XII

Навигаторские о-ва
Navigators Isl.

20.XII

круг Козерога
Capricorn

11.VIII

160°

180°

Courtesy of Technical University Munich.

CONTACT
Nikita Slavin
nslavin@gmail.com

SOFTWARE
ArcGIS Pro,
Adobe Illustrator

DATA SOURCES
Quantarctica

ROMANIAN TOPOGRAPHIC MAP, 1:50,000

National Cartographic Center of Romania
Bucharest, Romania
By the National Cartographic Center of Romania (CNC)

CONTACT
Radu Pantan
radu.pantan@cngcft.ro

SOFTWARE
ArcGIS Desktop 10.1

DATA SOURCES
National Cartographic Center of Romania (CNC)

The National Cartographic Center, the responsible authority for producing the official maps of Romania, initiated this project to address the need for an up-to-date map at a reasonable scale, which can be used as decisional support. The objective of the project was reached by achieving 737 printable map sheets at 1:50,000 scale, covering the entire territory of Romania.

The project successfully fulfilled its main objective, as well as obtaining the cartographic database at a scale of 1:50,000, which can be maintained and permanently updated, and exploring the possibility of offering and viewing web services. In this context, new concepts were introduced, such as cartographic database, model generalization, and graphic generalization. The development was possible due to the constant sustained effort for geospatial data collection.

The map sheets were automatically generated using MPS Atlas and include dynamic components such as nomenclature, location of the map sheet, slope guide, and the magnetic north declination.

Courtesy of CNC.

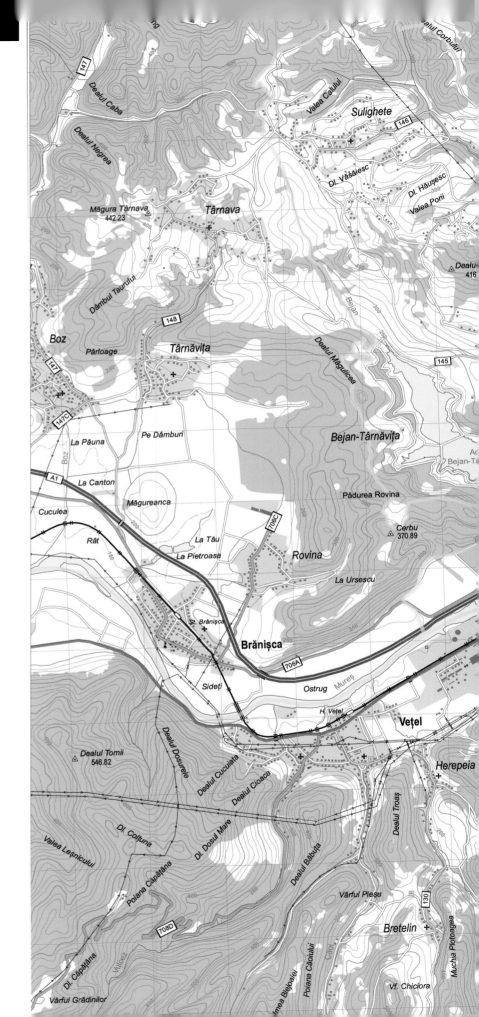

HYDRATING LANDSCAPES

Pearl River Eco Design, Poplarville, Mississippi, USA
By Ben Missimer

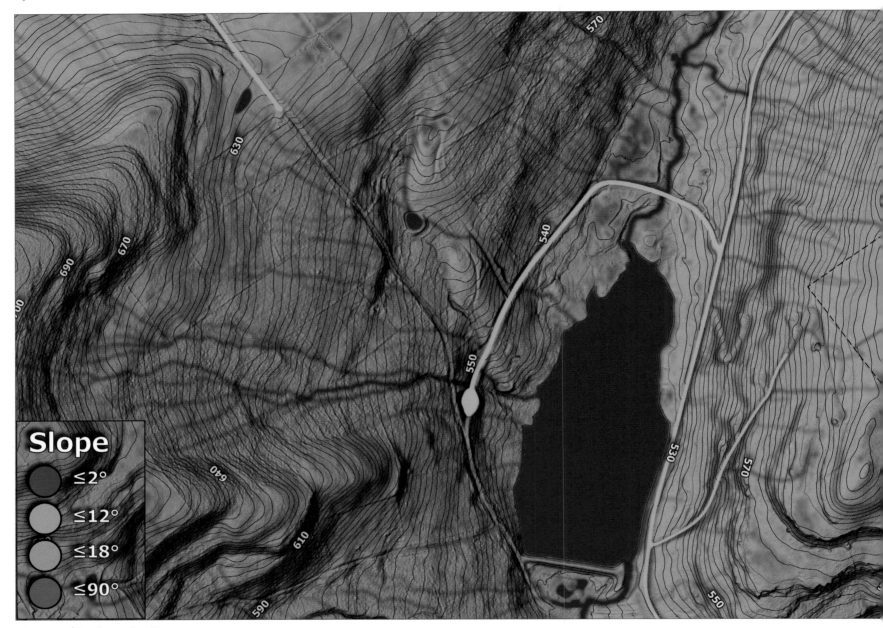

Courtesy of Pearl River Eco Design.

This colored elevation map depicts the altitude range of a location. It also shows how water flows across the landscape and where it collects in human-made ponds. This is useful when planning water retention strategies. Identifying key areas high on the landscape where we can hold water is a great benefit. Those water bodies would allow nature to hydrate and create a source of pressurized water for tree, crop, and animal needs below. Additional earthworks are planned, such as swales, spillways, berms, terraces, and ditches, which will slow, spread, and retain large quantities of water. Pacifying and retaining water versus allowing it to freely shed off the landscape not only passively irrigates our food but also recharges our rapidly depleting aquifers. These actions are critical for erosion control and soil loss prevention. This site will someday host an ecovillage that focuses on regenerative agriculture.

The slope map rapidly identifies targeted ranges of slope degrees. The 0–2° range represents low

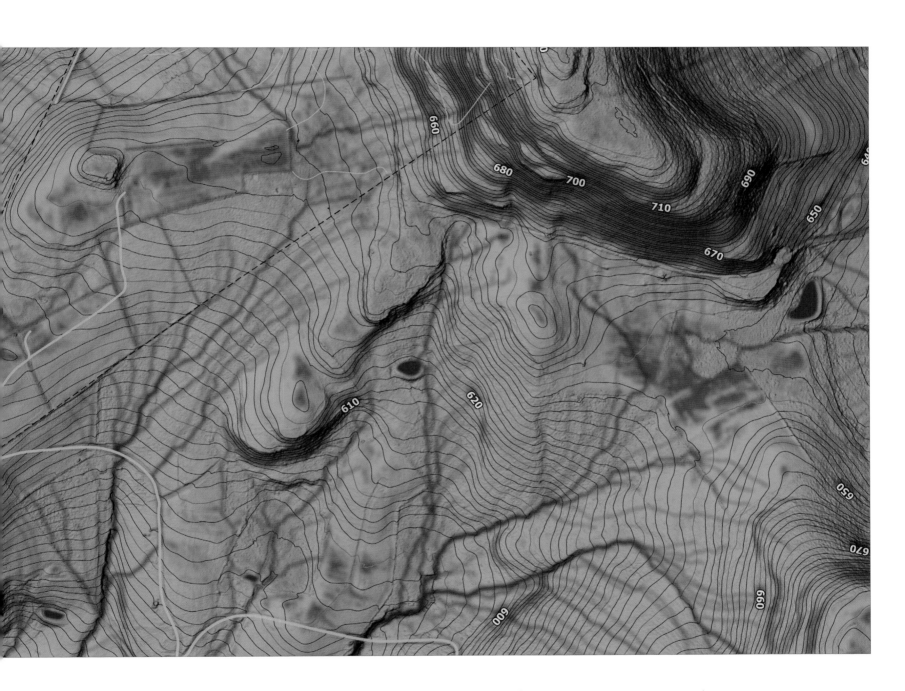

erosion zones and is ideal for garden plots. The 2–12° range represents where earthworks are most appropriate—anything beyond that can be a tip hazard for machinery. The 12–18° range represents areas most suitable for productive forestry, and the 18-90° range represents high erosion potential areas. It's best to leave those areas for wildlife.

CONTACT

Ben Missimer
ben@pearlriverdesign.com

SOFTWARE

ArcGIS Pro, Adobe Illustrator, Adobe Photoshop

DATA SOURCES

Citation: USGS Lidar Point Cloud NY MadisonOtsegoCo 2015 5190069650 LAS 2017

PRESERVING THE CRADLE OF SOUTHERN APPALACHIA

University of Tennessee at Chattanooga
Chattanooga, Tennessee, USA
By Charlie Mix

CONTACT

Charlie Mix
charles-mix@utc.edu

SOFTWARE

ArcGIS Pro, Adobe Illustrator

DATA SOURCES

Esri Green Infrastructure, The Nature Conservancy Resilient
& Connected Landscapes, USGS Protected Areas Database,
National Conservation Easement Database, National
Hydrography Dataset

The tri-state region surrounding Chattanooga, Tennessee, is
one of the most biodiverse and least protected regions in
the United States, encompassing the Cumberland Plateau,
the Ridge and Valley, and the Southern Blue Ridge Mountain
escarpments of Appalachia. This map represents conservation
priorities for Thrive Regional Partnership's Natural Treasures
Alliance, a conservation partnership of more than 30
organizations in the region.

Through the Cradle of Southern Appalachia initiative,
these partners work collectively to accelerate the pace of
conservation in the greater Chattanooga, Tennessee, region,
focusing on the high-priority areas identified in the map. The
conservation priority model was developed by overlaying
data reflecting climate resilience, intact habitat cores, habitat
fragments, wildlife corridors, and proximity to other protected
areas. Specific attributes and weighting were determined by
a committee of conservation leaders actively working in the
region.

A critical prioritization tool, this map and conservation
suitability model supports collaborative conservation projects
that support biodiversity, protect viewsheds, identify areas
suitable for sustainable outdoor recreation, and preserve a
cultural legacy of Appalachia that is inextricably linked to the
landscape.

Courtesy of University of Tennessee at Chattanooga.

Dark green represents areas that contain the highest quality lands with unfragmented and connected habitat, which offers natural systems resilience to climate change. Lower scoring areas still might contain rare and threatened species and ecosystems or hold other conservation values.

Lowest
Highest
Low
High
Medium

ProtectedLands

NORTHERN GREAT PLAINS

GreenInfo Network
Oakland, California, USA
By Maegan E. Leslie Torres

CONTACT
Maegan E. Leslie Torres
gin@greeninfo.org

SOFTWARE
ArcGIS Pro 2.0

DATA SOURCES
The Nature Conservancy, PAD-US; photos: Richard Hamilton Smith, Kenton Rowe, Dave Hanna, Amanda Hefner, Christ Helzer, Christopher Beltz, Jim Gores

The Nature Conservancy (TNC) has a big, landscape-scale story to tell about its work with Native tribes and ranchers across the prairies of the Northern Great Plains. We made a series of six showcase maps of the work, including three thematic views of the whole region and three highlights of specific TNC preserves. This project presented a challenge to incorporate complex messaging and vastly different map scales across several maps. We wanted these maps to have a clear family resemblance, while also each presenting its distinct message. We built the maps entirely in ArcGIS Pro, including setting all types, iconography, and inset photos.

Courtesy of GreenInfo Network.

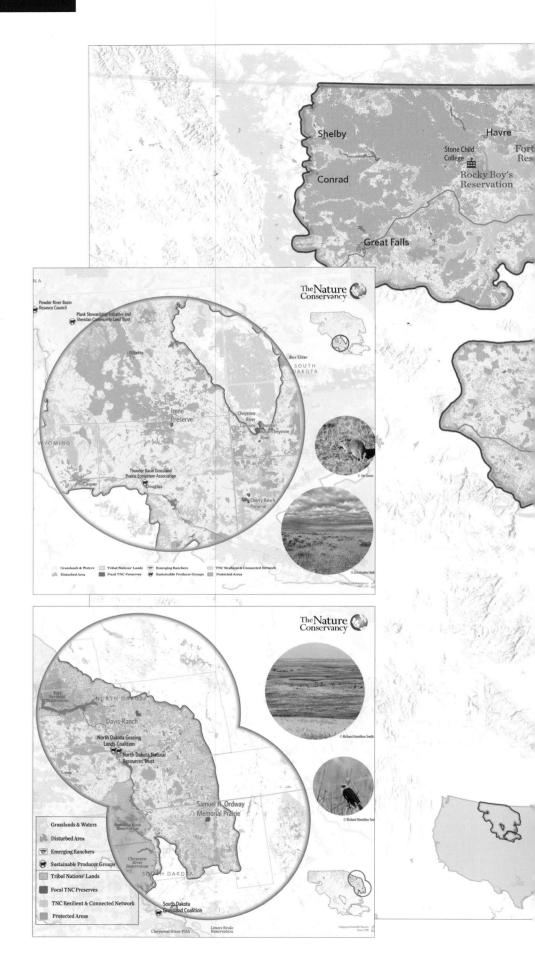

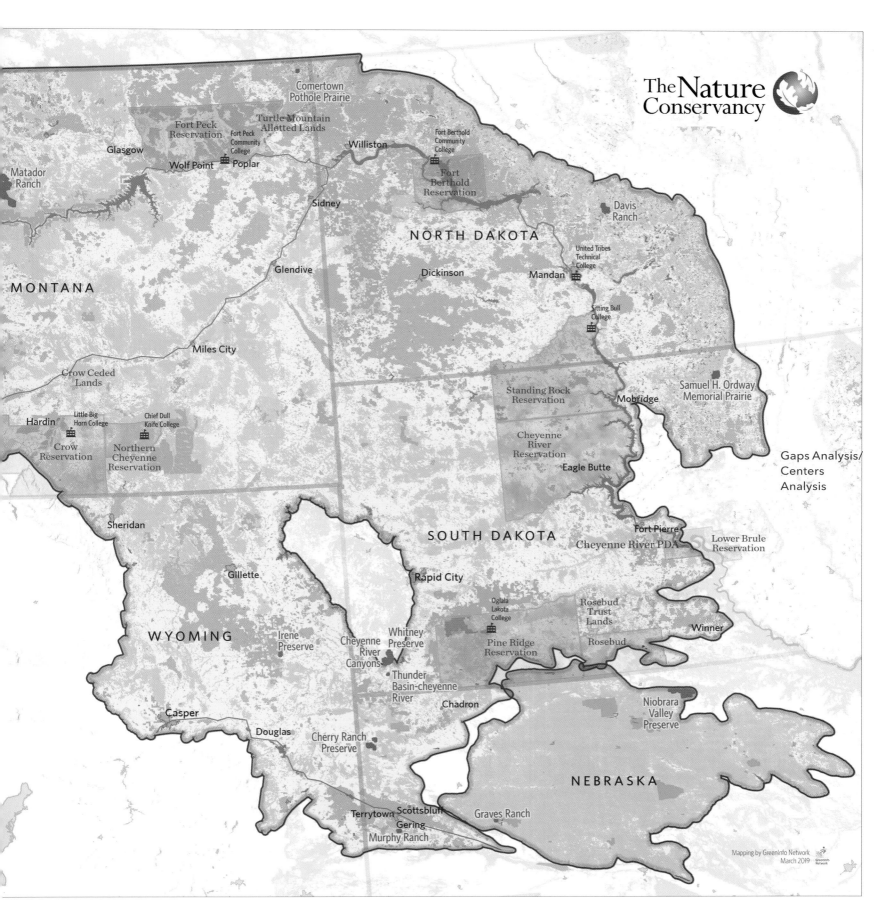

The Nature Conservancy

MONTANA

NORTH DAKOTA

SOUTH DAKOTA

WYOMING

NEBRASKA

Matador Ranch

Glasgow

Comertown Pothole Prairie

Fort Peck Reservation

Fort Peck Community College

Turtle Mountain Allotted Lands

Williston

Fort Berthold Community College

Wolf Point

Poplar

Sidney

Fort Berthold Reservation

Davis Ranch

Glendive

Dickinson

Mandan

United Tribes Technical College

Miles City

Sitting Bull College

Crow Ceded Lands

Standing Rock Reservation

Mobridge

Samuel H. Ordway Memorial Prairie

Hardin

Little Big Horn College

Chief Dull Knife College

Crow Reservation

Northern Cheyenne Reservation

Cheyenne River Reservation

Eagle Butte

Gaps Analysis/ Centers Analysis

Sheridan

Fort Pierre

Lower Brule Reservation

Cheyenne River PDA

Gillette

Rapid City

Oglala Lakota College

Rosebud Trust Lands

Winner

Irene Preserve

Cheyenne River Canyons

Whitney Preserve

Pine Ridge Reservation

Rosebud

Casper

Thunder Basin-cheyenne River

Chadron

Niobrara Valley Preserve

Douglas

Cherry Ranch Preserve

Graves Ranch

Terrytown

Scottsbluff

Gering

Murphy Ranch

Mapping by GreenInfo Network
March 2019

17

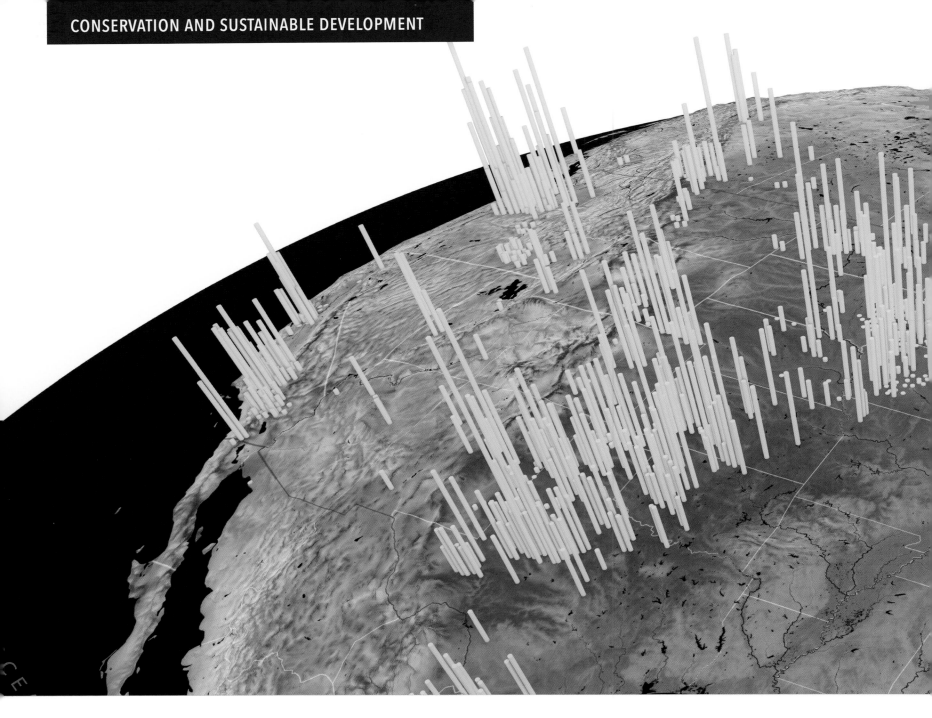

WHERE THE WIND BLOWS

Woodwell Climate Research Center, Falmouth, Massachusetts, USA
By Carl Churchill

Wind energy infrastructure relies on readily available supplies of wind power. In the United States and Canada, wind energy is generally strongest in the Great Plains region and around the Great Lakes. Politics also affects the placement of wind energy infrastructure. Infrastructure in this context means wind turbine farms, though they are not all equal. Total energy capacity is one variable for interpreting the scale of a wind turbine farm.

It can be hard to conceptualize both the distribution of available wind energy (in the form of wind speeds) and the concentration of wind energy infrastructure on a national scale. By providing a zoomed-out view of the bulk of North America, this map helps display the relationships between geography, politics, and local wind energy infrastructure in a single static view. This map also serves as an example of how 3D can be used to show an extra dimension of data (in this case, height) to add more context for readers.

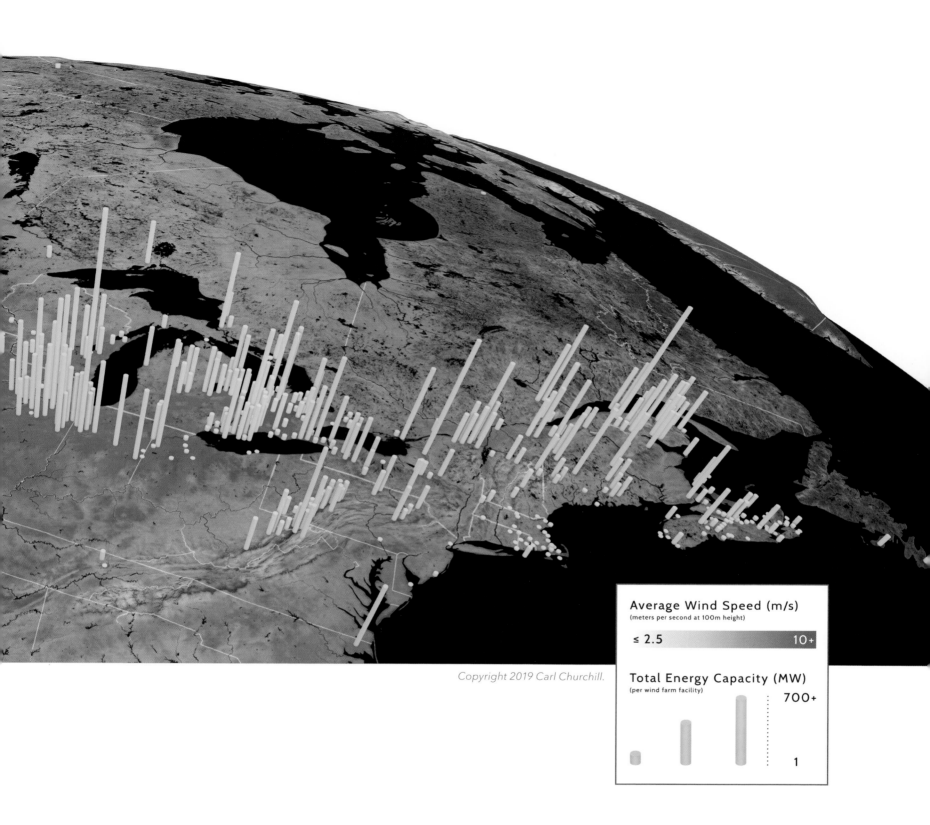

Average Wind Speed (m/s)
(meters per second at 100m height)

≤ 2.5   10+

Total Energy Capacity (MW)
(per wind farm facility)
                            700+

                              1

Copyright 2019 Carl Churchill.

CONTACT
Carl Churchill
cchurchill863@gmail.com

SOFTWARE
ArcGIS Pro, Adobe
Photoshop

DATA SOURCES
Natural Earth, Global Wind Atlas 3.0,
North American Cooperation on Energy
Infrastructure (NACEI), HydroSHEDs

ECOLOGICAL CORRIDORS: A NATURE-BASED CLIMATE SOLUTION

Nature Conservancy of Canada / Conservation de la nature Canada, Québec, Canada
By Maude Benny-Dumont

CONTACT
Kateri Monticone
kateri.monticone@natureconservancy.ca

SOFTWARE
ArcGIS Online, ArcGIS StoryMaps℠, ArcGIS Server, ArcMap™

DATA SOURCES
Conservation de la nature Canada (CNC) / Nature Conservancy of Canada (NCC); Réseau de milieux naturels protégés; Government of Québec (MELCC, MERN, MFFP); Government of Canada (ECCC); The Nature Conservancy (TNC); New Brunswick Natural Resources and Energy Development; Nova Scotia Environment; Nova Scotia Department of Natural Resources; The Department of Environment, Energy and Climate Action of Prince Edward Island; Ontario Ministry of Natural Resources and Forestry; Central New Hampshire Regional Planning Commission; Vermont Agency of Natural Resources; Maine Department of Agriculture, Conservation and Forestry.

The Ecological Corridors project, targeting natural passageways through which wildlife can move from one habitat to another, is helping to bring everyone together to counter the effects of climate change on biodiversity and our well-being. In collaboration with more than 100 experts and stakeholders, this initiative aims to mobilize the population, as well as key players in the protection, restoration, and sustainable development of ecological corridors throughout southern Québec.

Two complementary strategies are being developed to deal with climate change: mitigation and adaptation, in line with the priorities of the Government of Québec and the Government of Canada.

Courtesy of Conservation de la nature Canada / Nature Conservancy of Canada.

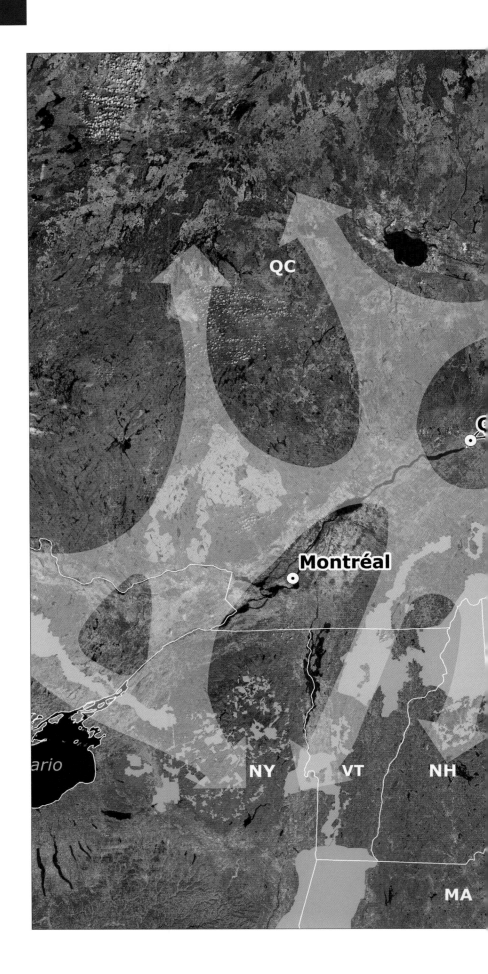

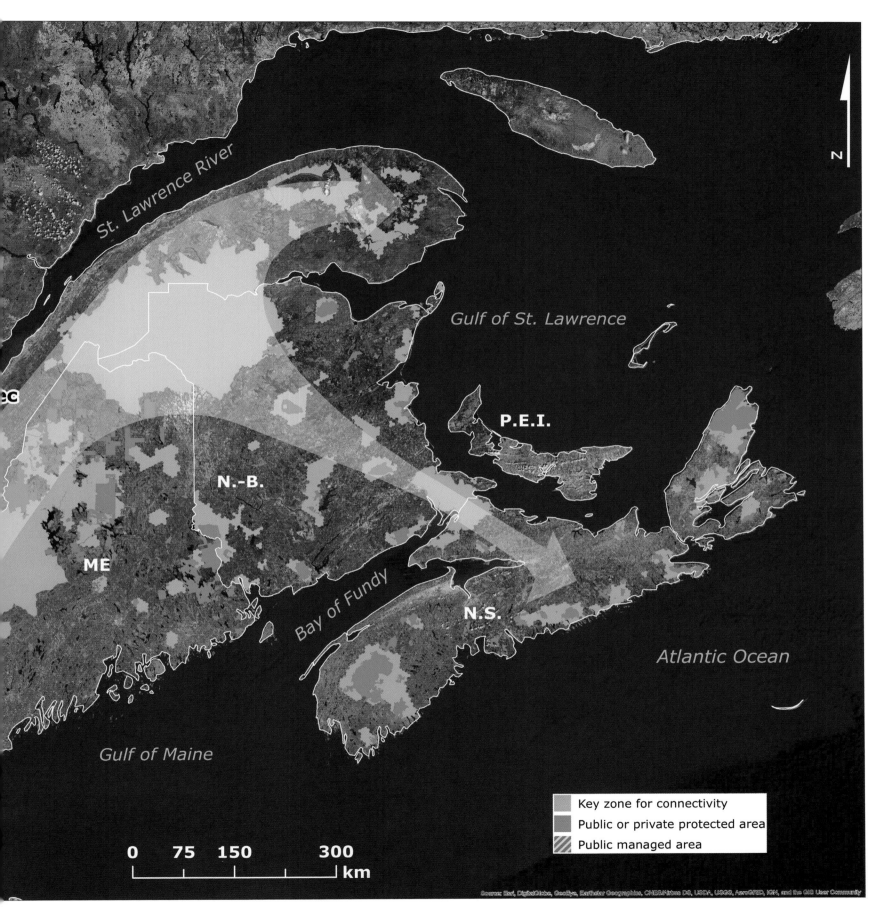

St. Lakrence River

Gulf of St. Lawrence

P.E.I.

EC

N.-B.

ME

Bay of Fundy

N.S.

Atlantic Ocean

Gulf of Maine

	Key zone for connectivity
	Public or private protected area
	Public managed area

0   75   150   300
km

RAPID ASSESSMENT OF NEW CONSERVATION SCIENCE IN THE NORTHERN APPALACHIAN– ACADIAN ECOREGION

Nature Conservancy of Canada / Conservation de la nature Canada, Québec, Canada
By Maude Benny-Dumont

CONTACT
Maude Benny
maude.bennydumont@natureconservancy.ca

SOFTWARE
ArcGIS Desktop 10.6.1

DATA SOURCES
Priority Location Site (2008) by Two Countries One Forest, Resilient and Connected Landscapes (2016) by The Nature Conservancy, Linkage Areas by Staying Connected Initiative (SCI)

This map depicts the overlaps between two studies of conservation priorities in the Northern Appalachian-Acadian (NAPA) Ecoregion. The first study, Priority Locations for Conservation Action, was published by Two Countries, One Forest (2C1Forest) in 2008. The second, Resilient and Connected Landscapes for Terrestrial Conservation, was carried out by The Nature Conservancy (TNC) and was completed in 2016. The 2C1Forest-only priority locations are shown in yellow, and the TNC-only areas are shown in blue. The locations where they overlap are shown in light and dark green. High threat zones are dark green, while the brighter green indicates relatively lower threat, as determined by the 2C1Forest study.

About 30% of the ecoregion, more than 26 million acres (nearly 11 million hectares), is identified as important by both studies. These include areas such as the Adirondacks, mountainous regions of Vermont and New Hampshire, northern and "downeast" Maine, northern and coastal New Brunswick, the Gaspé Peninsula, and large portions of Nova Scotia. The boundary of the NAPA ecoregion is shown as are the boundaries of nine landscape linkages identified by the Staying Connected Initiative.

Courtesy of Two Countries One Forest and Nature Conservancy of Canada.

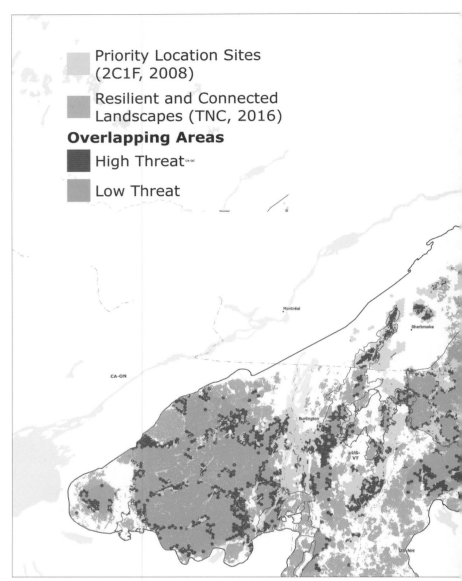

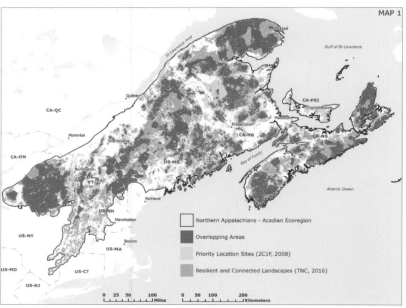

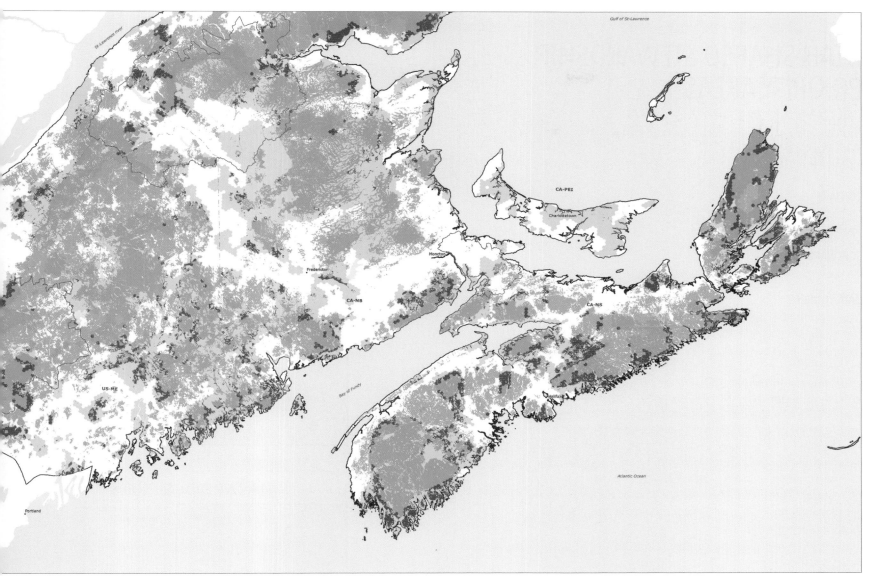

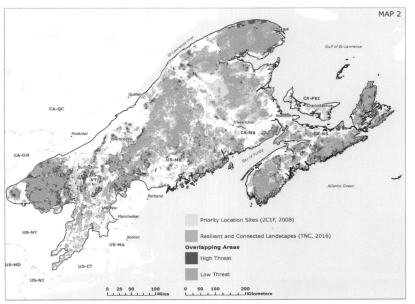

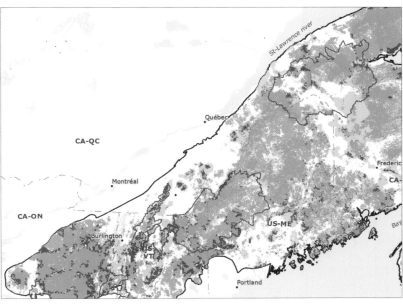

MAP 2

Priority Location Sites (2C1F, 2008)

Resilient and Connected Landscapes (TNC, 2016)

Overlapping Areas

High Threat

Low Threat

UTAH SHARED STEWARDSHIP PRIORITY AREAS

Utah Division of Forestry, Fire and State Lands
Salt Lake City, Utah, USA
By Michelle Baragona

CONTACT
Michelle Baragona
michelleb@utah.gov

SOFTWARE
ArcGIS Pro 2.4

DATA SOURCES
Utah Division of Forestry, Fire and State Lands;
Tim Metzger et al. (USFS)

Shared stewardship is a cooperative approach to managing Utah's forests. It enables land managers to confront the urgent forest health challenges that no single agency can face by itself. Utah's shared stewardship agreement provides a framework for the State of Utah and the US Forest Service to work together to identify forest health priorities that focus on restoration projects. The primary goals of the projects are protecting communities and watersheds from the threat of large unwanted wildfires.

A team of analysts from the US Forest Service and the Utah Division of Forestry, Fire and State Lands used three criteria to determine Utah's shared stewardship priority landscapes. Drinking water, strategic protection areas (in other words, values at risk), and fuel hazards were used to identify where the threat of fire presents the greatest risk to Utah's communities and water resources. This statewide assessment will be used by foresters, fire and fuels experts, and other natural resource specialists to design projects that will provide meaningful results in wildfire risk reduction in areas that directly affect people in the state of Utah.

Courtesy of Utah Division of Forestry, Fire and State Lands and the United States Forest Service.

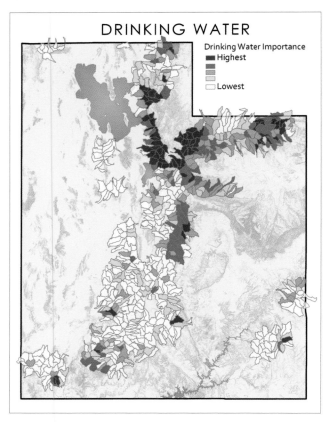

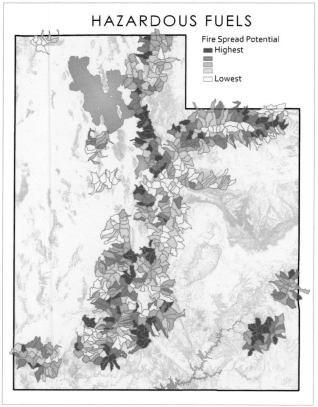

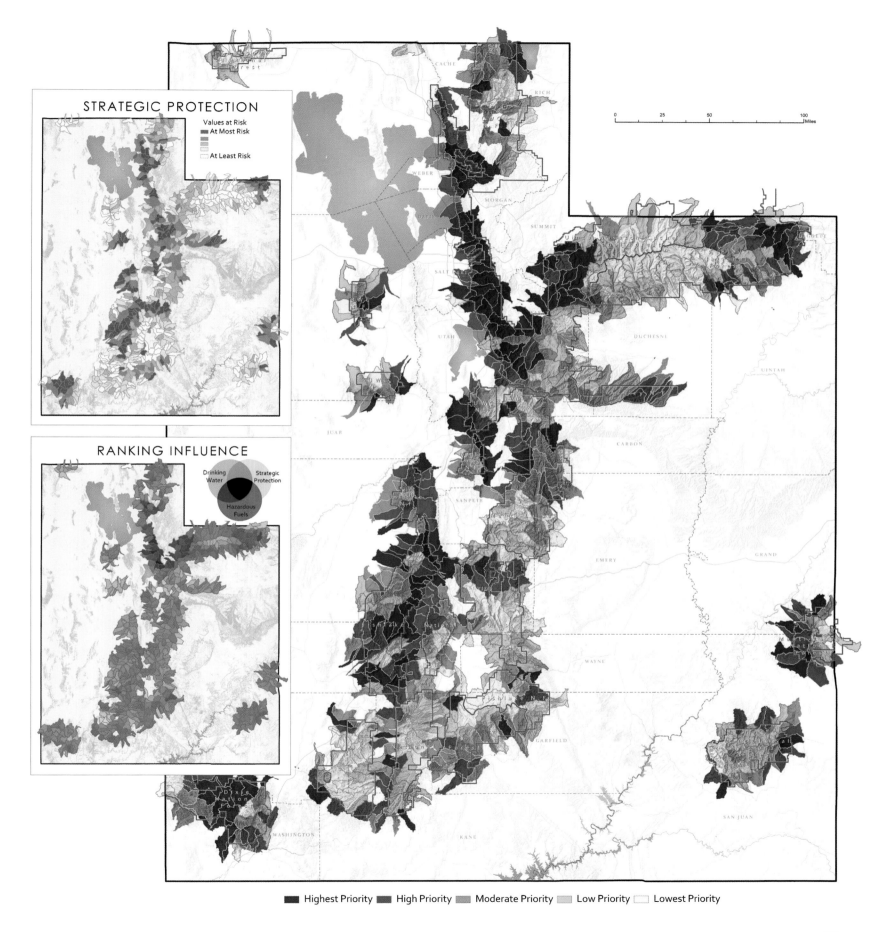

STRATEGIC PROTECTION

Values at Risk

At Most Risk

At Least Risk

RANKING INFLUENCE

Drinking Water

Strategic Protection

Hazardous Fuels

■ Highest Priority ■ High Priority ■ Moderate Priority Low Priority □ Lowest Priority

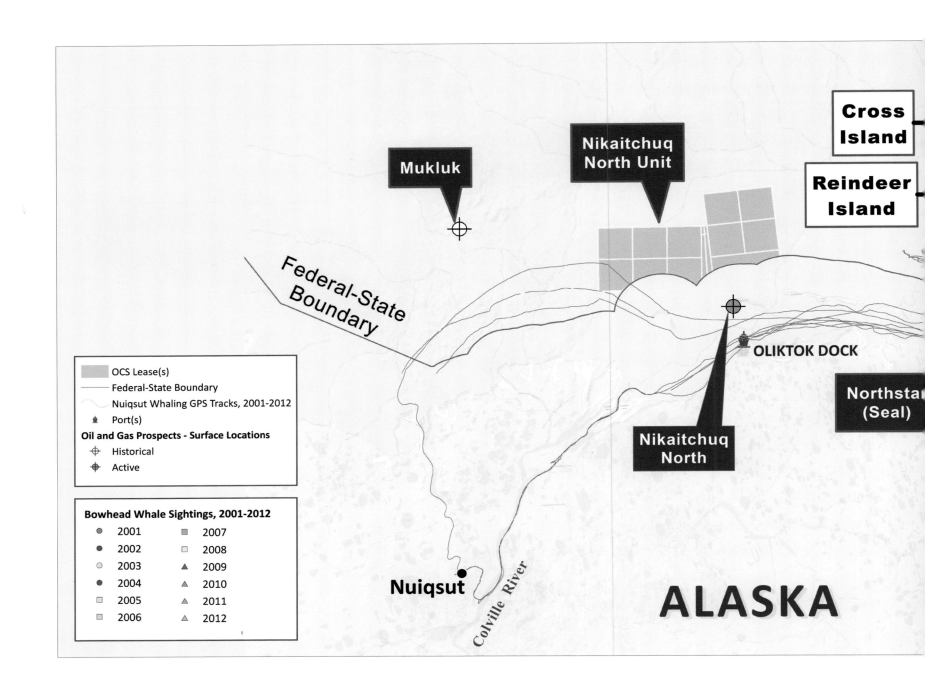

WHALING VESSEL TRACKS AND REPORTED WHALE SIGHTINGS FOR CROSS ISLAND

Bureau of Ocean Energy Management, Alaska Office, Anchorage, Alaska, USA
By Carla Langley and James Lima

Iñupiaq crews from the inland community of Nuiqsut, Alaska, conduct their annual bowhead whale harvest from a seasonal shore base on Cross Island in the Beaufort Sea in the US Arctic. The community's traditional knowledge emphasizes the cultural importance of the bowhead whale harvest. One major concern expressed by the whalers is the

potential of offshore oil and gas activities to deflect bowheads away from the Cross Island area, preventing a successful harvest. As part of a Bureau of Ocean Energy Management (BOEM)-sponsored study, the crews, equipped with handheld GPS units, recorded track lines of boats as they scouted for whales and locations where crews sighted whales for each year from 2001 to

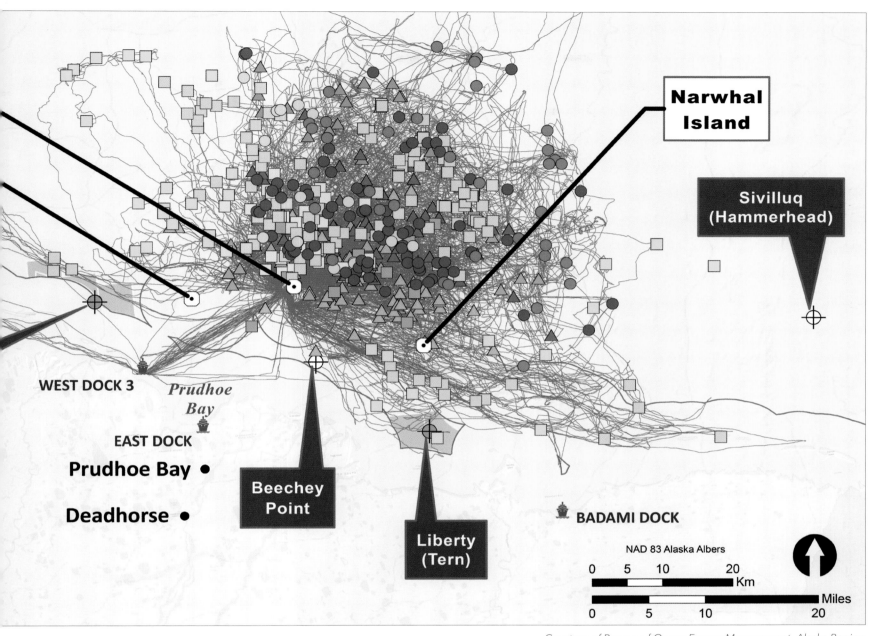

Narwhal Island

Sivilluq (Hammerhead)

WEST DOCK 3

Prudhoe Bay

EAST DOCK

Prudhoe Bay ●

Deadhorse ●

Beechey Point

BADAMI DOCK

Liberty (Tern)

NAD 83 Alaska Albers

| 0 | 5 | 10 | 20 |
Km

| 0 | 5 | 10 | 20 |
Miles

Courtesy of Bureau of Ocean Energy Management, Alaska Region.

2012. Compiled and mapped annually, the resulting composite map shows the cumulative geographic extent of the harvest area. Using this geospatial data and other information, BOEM develops spatial and temporal limits on offshore energy activities until the Nuiqsut bowhead whale quota has been met or until the Cross Island whalers officially end their whaling activities for the season.

CONTACT
Carla Langley
carla.langley@boem.gov

SOFTWARE
ArcGIS Desktop 10.4

DATA SOURCES
BOEM Alaska: O&G Prospects, Facilities, Cadastre (OCS Blocks, SLA boundary), Whaling tracks/sightings; USGS: Place Names (towns/villages)

AFRICAN RAPTOR DATABANK: EGYPTIAN VULTURE

Rookwood Studios, Haverfordwest, Pembrokeshire,
United Kingdom
By R. Davies, M. Virani, D. Ogada, A. Botha, R. Buij, J. Brouwer,
C. Barlow, H. Azafzaf, C. Kendall, R. Watson, and C. McClure.

CONTACT
Rob Davies
rag@rookwoodstudios.co.uk

SOFTWARE
ArcGIS Server

DATA SOURCES
African Raptor Databank, African Impact, BirdLife International & NatureServe, BirdLife Bulgaria (BSPB), BirdLife Tunisia (AAO), Boise State University, CITES (MIKE Database), Endangered Wildlife Trust, Hawk Conservancy Trust, Hawk Mountain Sanctuary, International Union for Conservation of Nature (African Elephant Database & Redlist Maps), Israel Nature & Parks Authority, Natural History Museum (TRING), Movebank, Niokolo-Koba Citizen Science Project, NOé Conservation, North Carolina Zoo, Raptors Botswana, Rare and Endangered Species Trust, Royal Society for the Protection of Birds, Sahara Conservation Fund, San Diego Zoo, Tanzanian Bird Atlas, The Peregrine Fund, University of Utah, Vulpro, West African Bird Database, Wildlife Act, Wildlife Conservation Society, Zoological Society London

The African Raptor Databank (ARDB) was built by about 600 bird watchers across Africa who submitted their sightings via a custom mobile app to a central server managed by Habitat Info. The project benefited from the Esri grant program and discounted licensing was paid for by The Peregrine Fund. This map shows historical sightings, satellite tracking of movements, range distribution, and a habitat model for the endangered Egyptian vulture. Recent records are withheld on account of their highly sensitive nature. The ARDB uses such information from all raptor species to identify the healthiest and most intact areas of ecosystems on the continent. These don't always correspond to existing protected areas.

Courtesy of Habitat Info and The Peregrine Fund.

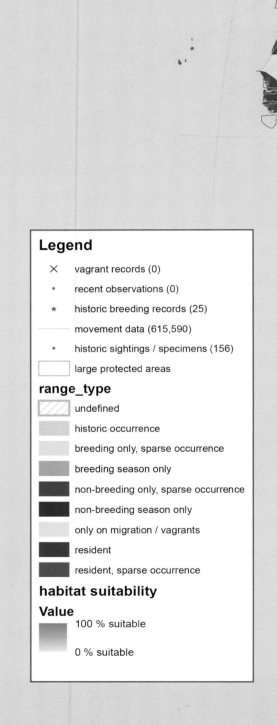

Legend

× vagrant records (0)

· recent observations (0)

★ historic breeding records (25)

— movement data (615,590)

· historic sightings / specimens (156)

☐ large protected areas

range_type

▨ undefined

▤ historic occurrence

breeding only, sparse occurrence

breeding season only

non-breeding only, sparse occurrence

non-breeding season only

only on migration / vagrants

resident

resident, sparse occurrence

habitat suitability

Value

100 % suitable

0 % suitable

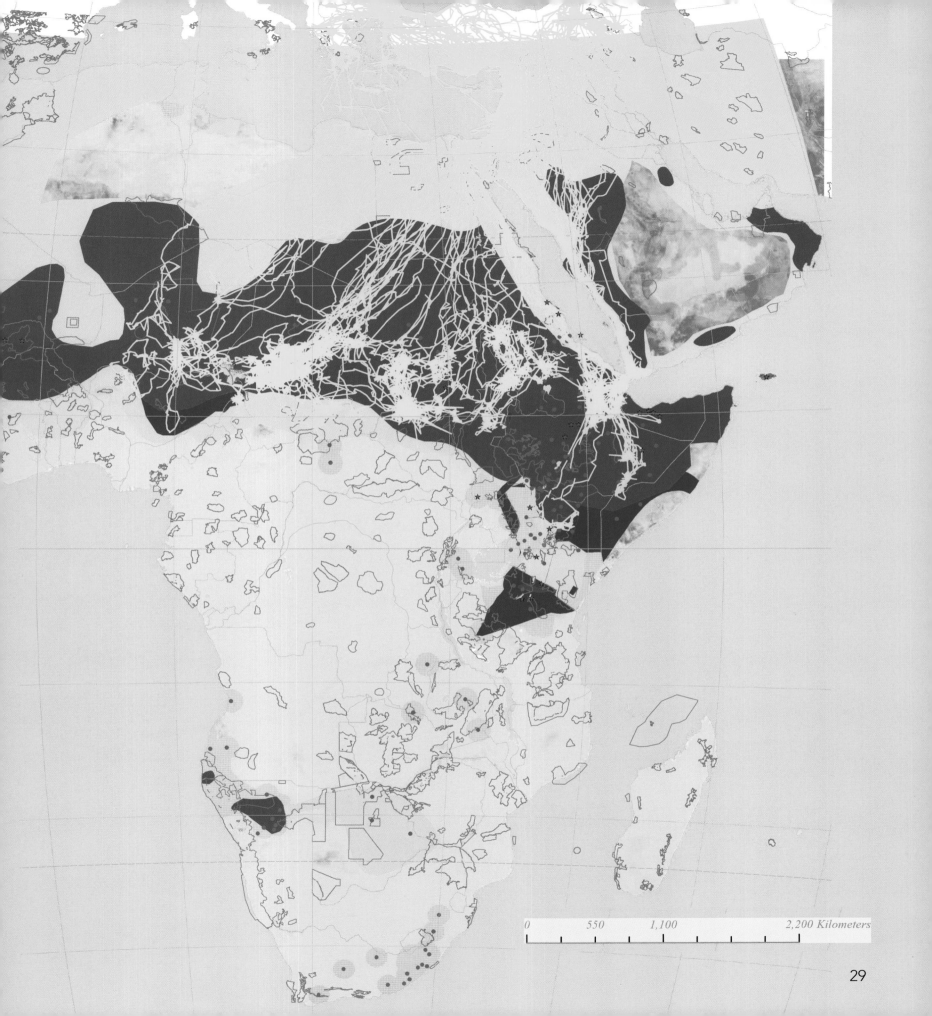

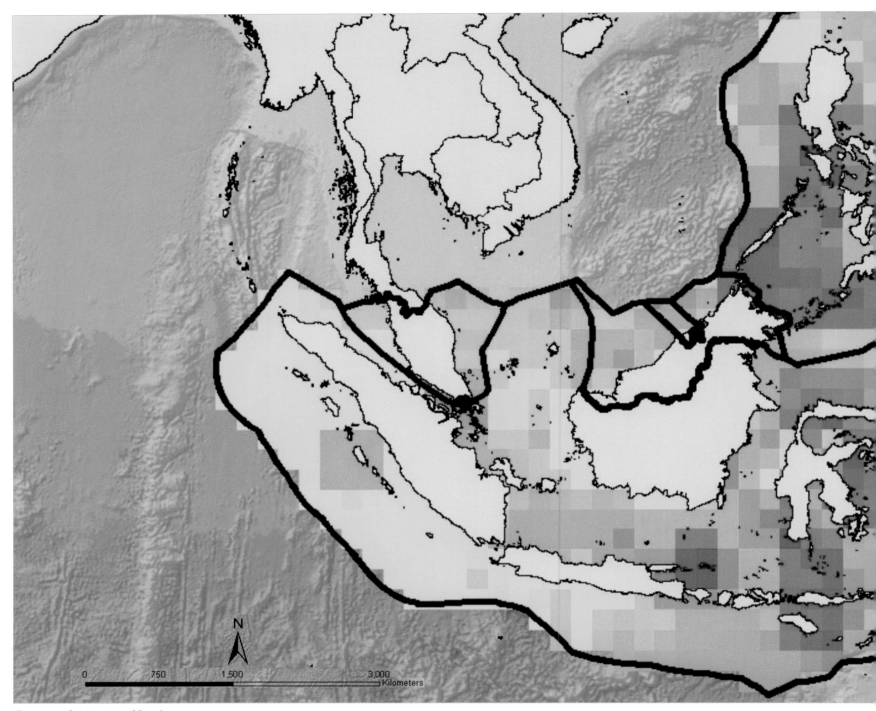

Courtesy of University of Southampton.

MULTICRITERIA EVALUATION OF CORAL BLEACHING, BIODIVERSITY, AND PLASTIC POLLUTION IN THE CORAL TRIANGLE

University of Southampton,
South Molton, Devon,
United Kingdom
By Amy Massey

This map displays a multicriteria evaluation of coral bleaching, biodiversity, and plastic pollution in the Coral Triangle in the western Pacific Ocean. Impact levels were rated from high to low in order to highlight which areas were at the highest risk. Viewers can see that reefs surrounding the Philippines are impacted the most by these factors.

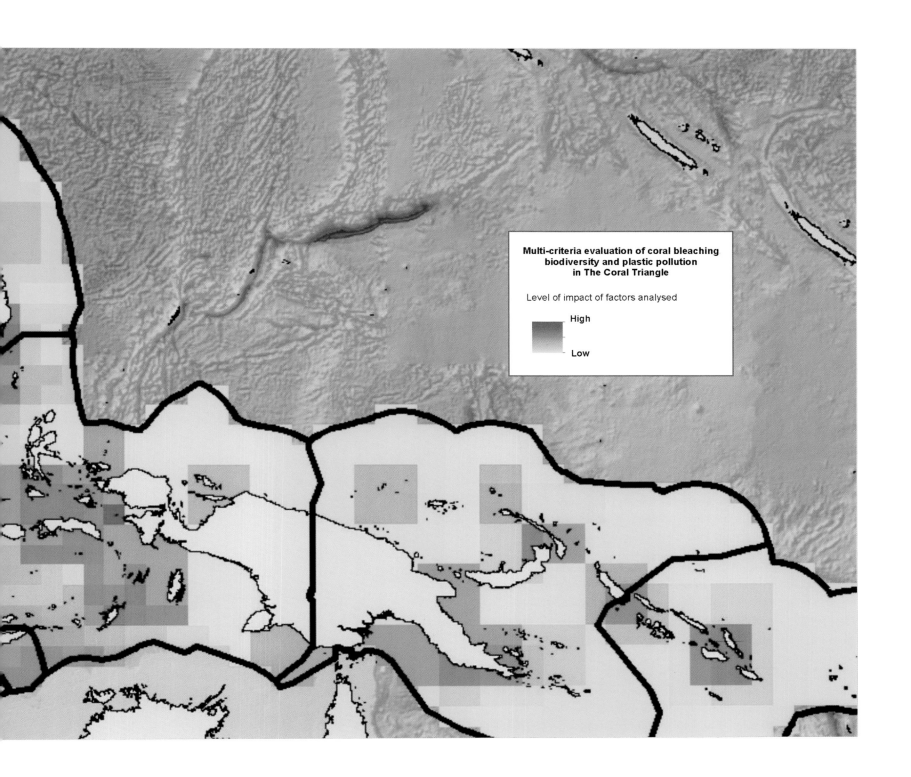

Multi-criteria evaluation of coral bleaching
biodiversity and plastic pollution
in The Coral Triangle

Level of impact of factors analysed

High

Low

CONTACT
Amy Massey
amymassey89@gmail.com

SOFTWARE
ArcGIS Desktop

DATA SOURCES
ReefBase (2019), WRI (2011), ArcGIS
Online (2019), Esri (2019), Sebille et al.
(2012), PlasticADrift.org (2019)

AUSTIN'S URBAN HEAT ISLAND

City of Austin
Austin, Texas, USA
By Alan Halter

CONTACT
Alan Halter
alan.halter@austintexas.gov

SOFTWARE
ArcGIS Pro 2.5

DATA SOURCES
Landsat-8 OLI/TIRS scenes 30-meter resolution (USGS/NASA)

In the US, heatwaves injure more people than all other natural disasters combined. So, the City of Austin is exploring ways to monitor urban heat to become more climate resilient. In a partnership of the City of Austin and Texas State University, students acquired seven Landsat-8 satellite images, from April to September 2014. They converted these images to land-surface temperature values using ArcGIS. First, they removed cloud cover obscuring the images. Then they averaged temperature data from all seven images resulting in the land-surface temperature data you see in this map. Red areas are hottest, corresponding with building densities and road pavement. Dark-blue areas are coolest, corresponding to waterways and tree canopy cover. Although more current imagery is available, the 2014 imagery coincides with available tree canopy data for that year.

Continuing this mapping effort, the City of Austin is a participating partner in the CDC and NOAA's 2020 Heat Watch Campaign to map temperatures in Austin using sensors. This will help supplement the land-surface temperature data by exploring ambient air temperature as one experiences heat walking down the street. For the campaign, data is collected by mounting a thermocouple device with a paired GPS unit on the passenger side of a car and driving the vehicle across east Austin—where vulnerability to heatwaves is strongest. These sample points are then interpolated to create an areawide predictive model based on the statistical relationships between land cover and temperature.

Courtesy of Texas State University's Department of Geography.

MLK

15th

State
Capitol

I-35

6th

Convention
Center

S 1st

Congress

Temperature

103°F

65°F

N

E

W

S

33

FROM DATA TO DECISIONS: APPLYING ARCGIS SURVEY123 AND PREDICTIVE MODELING TO GUIDE CONSERVATION

NatureServe
Arlington, Virginia, USA
By NatureServe

CONTACT
Samantha Belilty
Samantha_Belilty@natureserve.org

SOFTWARE
ArcGIS Survey123, ArcGIS Desktop

DATA SOURCES
NatureServe

Natural heritage biologists in NatureServe's network use ArcGIS Survey123 to collect standardized, georeferenced field data on our nation's imperiled plants and animals using mobile devices. That data is used to train predictive habitat models, such as the one shown here for the snuffbox mussel (Epioblasma triquetra), which is listed as endangered under the US Endangered Species Act. By analyzing climate, land cover, hydrology, and other environmental characteristics at these locations of documented occurrence (such as random forests) using spatial modeling techniques, it's possible to predict other locations on the landscape where a species is most likely to occur. This spatially explicit information enables the regulated community to more efficiently understand how their activities may, or may not, impact regulated species, while helping scientists and conservationists focus their management and protection efforts on areas where they're likely to have the greatest impact.

CONFLICT IN WEST AFRICA

Tesla Government Inc.
Vienna, Virginia, USA
By Tesla Government Inc.

CONTACT
Madison Marshall
gis@teslagovernment.com

SOFTWARE
ArcGIS Pro 2.5, Adobe Illustrator

DATA SOURCES
The Armed Conflict Location & Event Data Project (ACLED);
Department of State; United Nations Office for the
Coordination of Humanitarian Affairs (OCHA); The Global
Administrative Area Database (GADM)

This monthly product was created to display the distribution
of conflict events, including battles, explosions, riots, and
violence against civilians across 15 countries in West Africa.
Tesla Government Inc. uses the Armed Conflict Location
& Event Data (ACLED) project to track and visualize the
conflict events of areas of interest to enhance the situational
awareness of United States military and government
personnel, humanitarian aid organizations, and multinational
partners. This map is meant to provide insight into the high-
risk areas across the region and allow for the exploration of
conflict trends and shifts by providing a monthly update. The
availability of historical and current maps allows stakeholders
to build the most comprehensive understanding of the
region's conflict patterns to support their mission. The
thematic map was created using a 5,000-square-kilometer
tessellation and then by binning and symbolizing the
occurrence of events to specific geographic areas.

Courtesy of Tesla Government Inc.

Conflict Events
Count

≤1
≤2
≤5
≤10
>10

0   500 km

Projection=Africa Albers Equal Area Conic
Event Data: ACLED

OCEAN PLASTIC MAP

Technical University Munich
Munich, Germany
By Nikita Slavin

CONTACT

Nikita Slavin
nslavin@gmail.com

SOFTWARE

ArcGIS Pro, Adobe Illustrator

DATA SOURCES

ETOPO2 natural earth data; plastic input by rivers
(https://www.nature.com/articles/ncomms15611),
mismanaged plastic waste
(https://www.nature.com/articles/s41599-018-0212-7),
plastic particles concentration
(https://journals.plos.org/plosone/article)

A map—made entirely of plastic—showing areas of plastics
pollution of the oceans. More than 15 map layers are printed
on plastic foil. The stacked layers allow map users to "see
through" the ocean depths. The concentration of plastic
particles is shown via the density of symbols for an off-scale
depiction of typical waste items, including bottles, bags, and
brushes. Plastic debris and microplastics are transported by
ocean currents across international borders. Debris can be
found everywhere: on the most remote shores of uninhabited
islands, in the Arctic ice, on the ocean floor, and even inside
marine organisms. Environmental experts say that, while
our knowledge of the full impact of plastics in the oceans is
incomplete, what we already know indicates the necessity of
taking immediate action.

Courtesy of Technical University Munich.

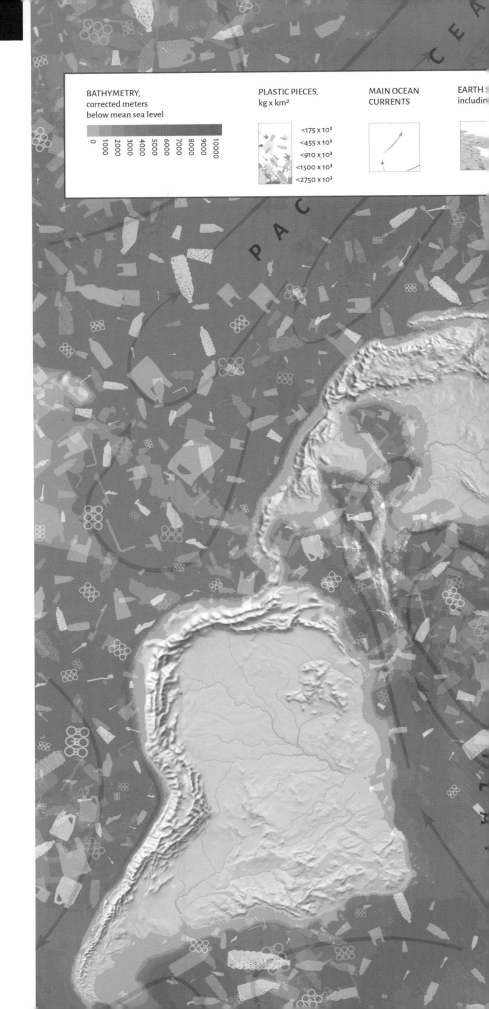

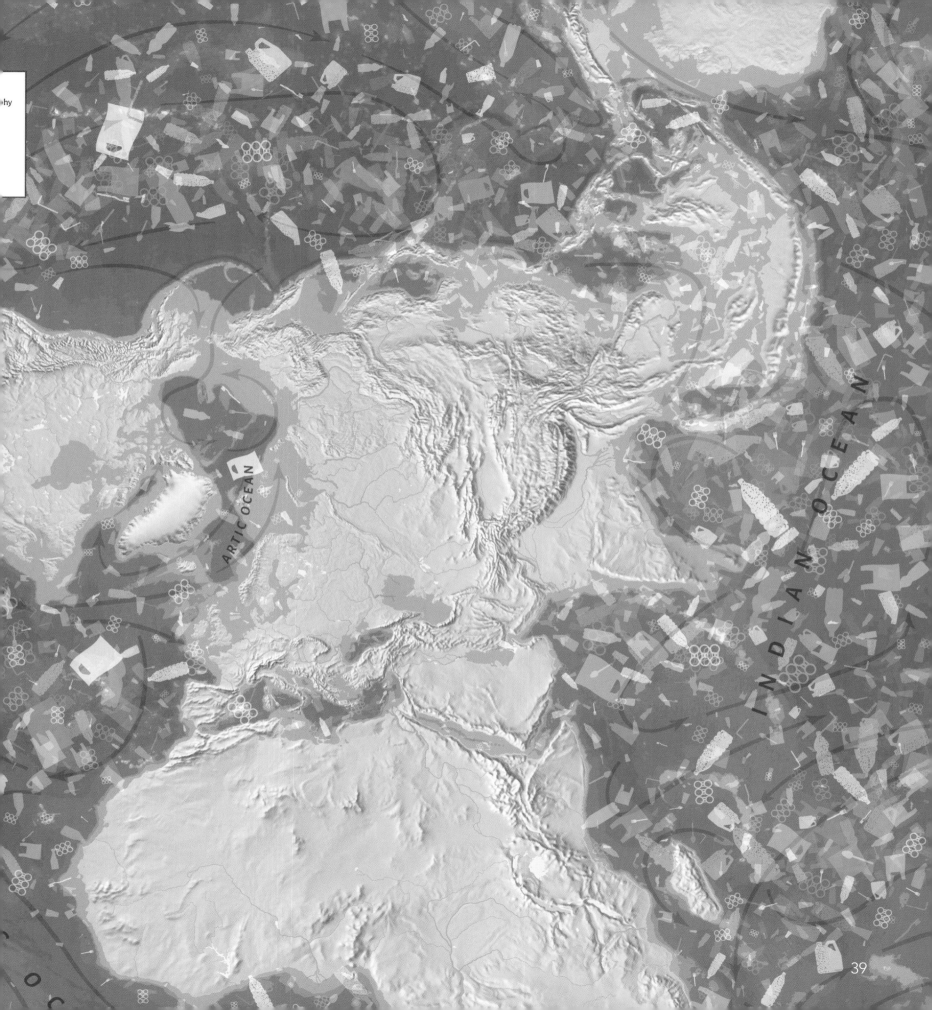

ARTIC OCEAN

INDIAN OCEAN

WISCONSIN INVASIVE SPECIES CONTROL

Stantec
Cottage Grove, Wisconsin, USA
By Jeremy Del Prete

CONTACT

Jeremy Del Prete
Jeremy.DelPrete@stantec.com

SOFTWARE

ArcGIS Online, ArcGIS StoryMaps, ArcGIS Pro 2.6

DATA SOURCES

Stantec, Wisconsin Department of Natural Resources (DNR), Wisconsin Department of Transportation, Lakeshore Natural Resource Partnership, Southeastern Wisconsin Invasive Species Consortium, University of Wisconsin–Green Bay, Ozaukee Washington Land Trust, Glacierland, Esri

Invasive plant species reduce biodiversity, which can cause an ecosystem to become more susceptible to environmental stress. Invasive species are often introduced through human activities, such as urban development, and spread quickly when unmanaged.

Successful treatment methods include a combination of herbicide application, mechanical mowing or cutting, prescribed fire, and, eventually, revegetation with native plants. To be effective, these treatments must be monitored and applied year after year, which can be difficult and expensive to coordinate at the scale necessary to make a meaningful difference.

We believe that gains in invasive plant management can be made through increased stakeholder collaboration to pool monetary and knowledge-based resources. To that effect, we have partnered with the Wisconsin DNR, Lakeshore Natural Resource Partnership, local landowners, and many others to create a management plan that involves surveying and tracking treatments of invasive species throughout Lake Michigan Basin counties over multiple years. To effectively communicate our progress within this large geographic extent, we created this map of our survey data aggregated into bins to generally show the areas we have treated so far, how many years each area has been treated, and the density of known invasive plant species. Highlighted is our early focus on key waterways near populated areas, which is where invasive species can be most prevalent, and thus where our efforts can do the most good.

Courtesy of Stantec.

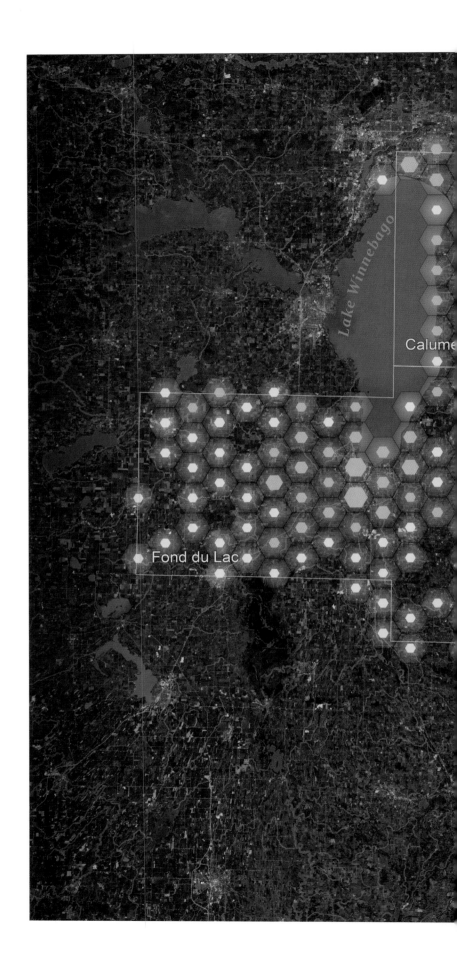

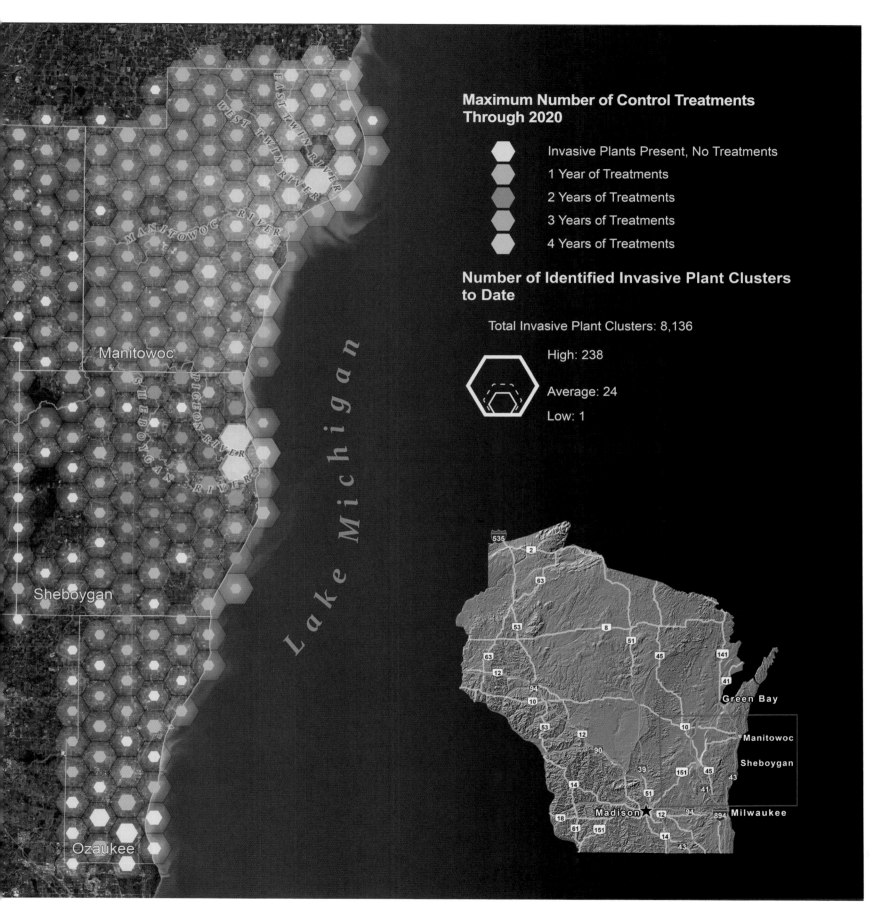

Maximum Number of Control Treatments Through 2020

Invasive Plants Present, No Treatments

1 Year of Treatments

2 Years of Treatments

3 Years of Treatments

4 Years of Treatments

Number of Identified Invasive Plant Clusters to Date

Total Invasive Plant Clusters: 8,136

High: 238

Average: 24

Low: 1

Lake Michigan

Manitowoc

Sheboygan

Ozaukee

EAST TWIN RIVER

WEST TWIN RIVER

MANITOWOC RIVER

SHEBOYGAN RIVER

PIGEON RIVER

Green Bay

Manitowoc

Sheboygan

Milwaukee

Madison

TEXAS WORKING LANDS

Texas A&M Natural Resources Institute
College Station, Texas, USA
By Garrett Powers and Alison Lund

CONTACT
Addie Smith
Addie.Smith@ag.tamu.edu

SOFTWARE
ArcGIS Pro

DATA SOURCES
Texas Comptroller of Public Accounts

At the Texas A&M Natural Resources Institute (NRI), we tell
the story of our state's privately owned farms, ranches, and
forests, otherwise known as working lands, which provide
numerous ecological, economic, and intrinsic benefits to our
communities and beyond. This map illustrates the patchwork
of working lands spread throughout the state. Each dot
represents an area of 10 acres of a specific agricultural land
use, distinguished by color. The resulting map provides a
visual impression of the varying agricultural landscapes that
define our state. Conversely, Texas A&M AgriLife supports
a network of research and extension centers to serve the
regional and local needs of private land stewards. NRI is
a unit of Texas A&M AgriLife, which works to improve and
sustain working lands through innovative research and
extension education. Thirteen Texas A&M AgriLife research
and extension centers exist across the state, each tailoring its
studies and outreach programs to provide relevant tools for
local landowners.

Courtesy of Texas A&M Natural Resources Institute.

● El Paso

TEXAS WORKING LANDS*

CROPLAND
TIMBER
WILDLIFE MANAGEMENT
ORCHARDS

1 DOT = 10 ACRES OF LAND

Cities labeled on map are home to Agrilife Research and Extension Centers
Data: Texas Comptroller of Public Accounts (Texas Property Tax Assistance Division)
*Grazing excluded; data summarized by independent school district (ISD)

Amarillo
Vernon
Lubbock
Dallas
Overton
Stephenville
San Angelo
Temple
Beaumont
Uvalde
Corpus
Christi
Weslaco

NOAA'S AMAPPS MARINE MAMMAL MODEL VIEWER

DoC/NOAA/NMFS/NEFSC
Woods Hole, Massachusettes, USA
By David F. Chevrier

CONTACT
David F. Chevrier
David.Chevrier@NOAA.gov

SOFTWARE
ArcGIS API for JavaScript

DATA SOURCES
NOAA, Esri

This map was created with ArcGIS API for JavaScript using data hosted on NOAA's ArcGIS Online organization. The Atlantic Marine Assessment Program for Protected Species (AMAPPS) is a comprehensive multiagency research program in the US Atlantic Ocean, from Maine to the Florida Keys. Its aims are to assess the abundance, distribution, ecology, and behavior of marine mammals, sea turtles, and seabirds throughout the US Atlantic and to place them in an ecosystem context. The AMAPPS Marine Mammal Habitat Model Viewer provides spatially explicit information in a format that can be used when making marine resource management decisions and provides enhanced data to managers by addressing data gaps that are essential to support conservation initiatives mandated under the Marine Mammal Protection Act (MMPA), the Endangered Species Act (ESA), and the National Environmental Policy Act (NEPA) (NMFS 2016). These seasonal spatially explicit density distribution maps were developed under phase I of the AMAPPS project using animal distribution data collected during shipboard and aerial line transect surveys during 2010–2014. Dive time data was derived from tagged and satellite and model-based static and dynamic environmental information.

Courtesy of DoC/NOAA/NMFS/NEFSC.

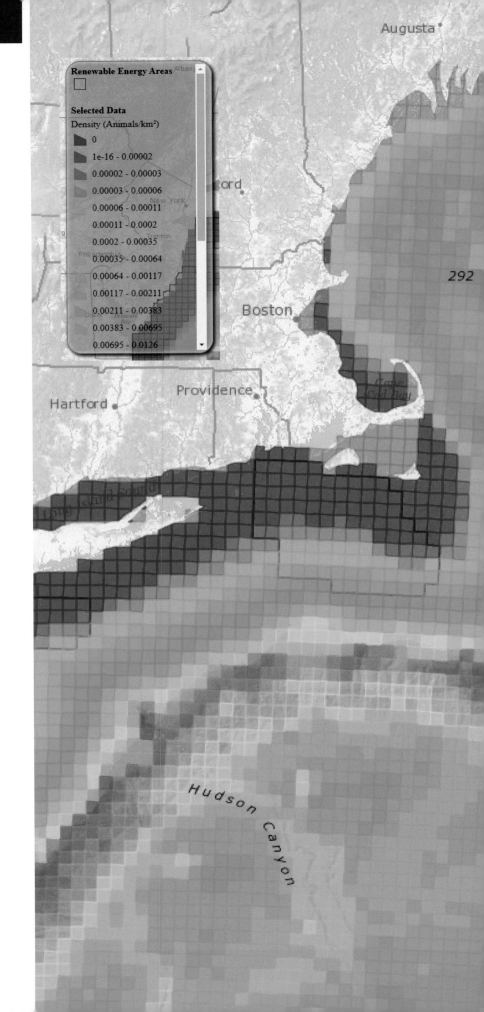

ADVOCACY FOR EQUITY IN NORTHEAST HOUSTON AFTER HURRICANE HARVEY

Rock Whisperer LLC
Tulsa, Oklahoma, USA
By Edith Newton Wilson

CONTACT

Edith Newton Wilson
edith@rockwhispererllc.com

SOFTWARE

ArcGIS Desktop

DATA SOURCES

US Census Bureau, Geography Division; OpenFEMA Dataset: Individual Assistance Housing Registrants - Large Disasters - v1; Harris County Flood Control District Education Mapping Tool Channels Data (all accessed in January or February 2019)

After Hurricane Harvey, residents of northeast Houston who experienced flooding partnered with community organizers and scientists to better advocate for flood relief. This map was one of a suite of maps provided to neighborhood leaders in support of those projects. It shows the average height of water in structures (inches) by census block in relation to Harris County Flood Control District (HCFCD) infrastructure, indicated by the blue lines representing ditches. Residents could easily identify their locations and determine which named HCFCD ditch or drainage required attention, as well as better present their case for equitable distribution of flood relief funds based on the severity of inundation.

Courtesy of Rock Whisperer LLC.

SCOTTSDALE FIRE INCIDENTS, 2006-2019

City of Scottsdale
Scottsdale, Arizona, USA
By Mele D. Koneya

CONTACT
Mele D. Koneya
mkoneya@scottsdaleaz.gov

SOFTWARE
ArcGIS Pro 2.6

DATA SOURCES
City of Scottsdale

On July 1, 2005, after more than 50 years of fire service having been provided by a private company, the City of Scottsdale formed the Scottsdale Fire Department (SFD) as a municipal department. In its first fiscal year of service, 2005-2006, SFD operated 13 stations previously used by the private provider. That year, these 13 stations responded to around 24,000 calls for service. By fiscal year 2019-2020, to better cover nearly 38,000 calls, SFD had expanded to 15 stations by adding stations 602 and 608. In addition to those two new facilities, two other stations, 601 and 613, were relocated to provide even better response times. In 2021, two more stations, 603 and 616, were moved to improve coverage within Scottsdale. This map, using 80-acre grid squares, illustrates the difference in SFD calls for service from 2005 to 2019.

Courtesy of City of Scottsdale.

MDA SITUATIONAL INDICATION LINKAGES (MSIL)

Japan Coast Guard, Chiyoda Ward, Tokyo, Japan
By Marine Spatial Information Service Office, Oceanographic Data and Information Division,
Hydrographic and Oceanographic Department, Japan Coast Guard

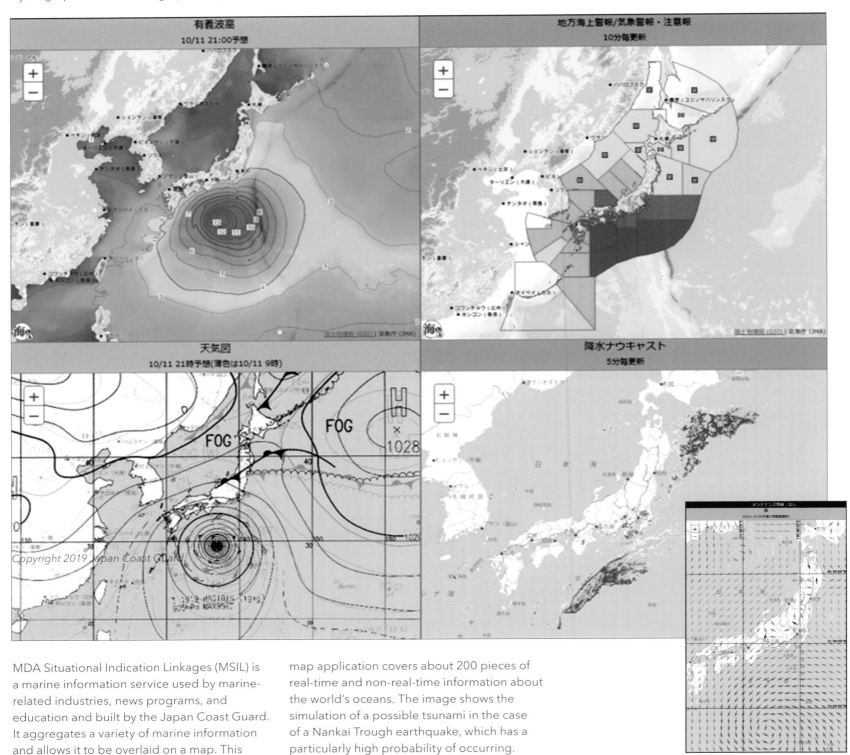

MDA Situational Indication Linkages (MSIL) is a marine information service used by marine-related industries, news programs, and education and built by the Japan Coast Guard. It aggregates a variety of marine information and allows it to be overlaid on a map. This map application covers about 200 pieces of real-time and non-real-time information about the world's oceans. The image shows the simulation of a possible tsunami in the case of a Nankai Trough earthquake, which has a particularly high probability of occurring.

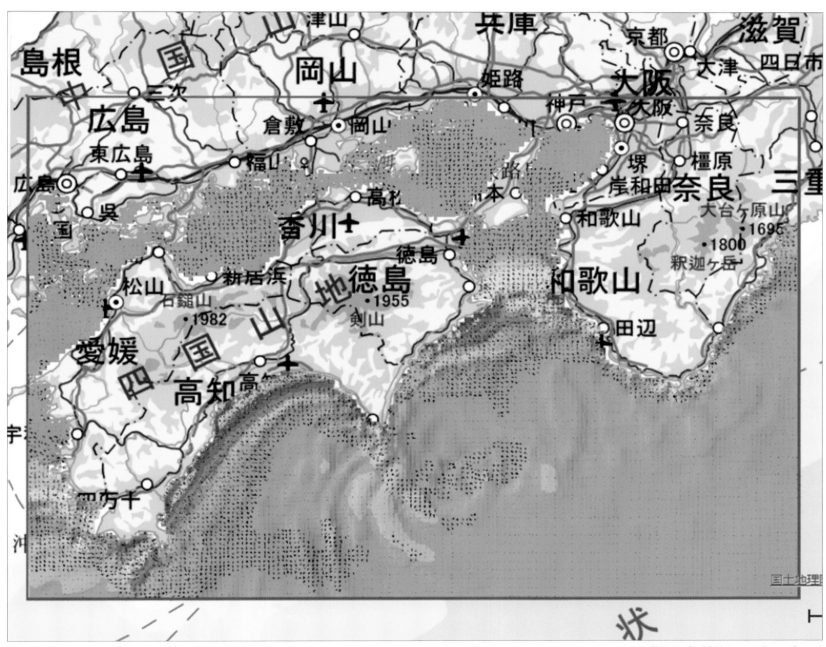

CONTACT
Hydrographic and Oceanographic
Department, kaiyogis@jodc.go.jp

SOFTWARE
ArcGIS Desktop,
ArcGIS Enterprise,
ArcGIS API for Javascript

DATA SOURCES
MDA Situational Indication Linkages (MSIL)

TOWN OF ERWIN SIDEWALK CONDITIONS

National 4-H GIS Leadership Team
Erwin, Tennessee, USA
By Christiana Giblert, Ben Merrit, Elisabeth Casey, Haley Russell, Logan Bennett, Joshua Gilbert, Rachel Bradshaw, Chris Mackey, Matthew Laws, Isabella Laws, Alaina Satterly, and Eli Cadwell

CONTACT
Christiana Gilbert
chrgil2003@gmail.com

SOFTWARE
ArcGIS Online, Microsoft PowerPoint

DATA SOURCES
ArcGIS Living Atlas of the World

The members of the National 4-H GIS Leadership Team decided we could help the most citizens by improving the sidewalks in residential areas. We noticed how citizens traveling by foot had a hard time getting around town, the elderly had a hard time with steeply sloped sidewalks, and most of the sidewalks had grass overtaking them, making it difficult for wheelchairs and strollers to get around. Because this is a struggle for many town residents, the team decided that this project would be useful to the majority of the public.

Courtesy of National 4-H GIS Leadership Team.

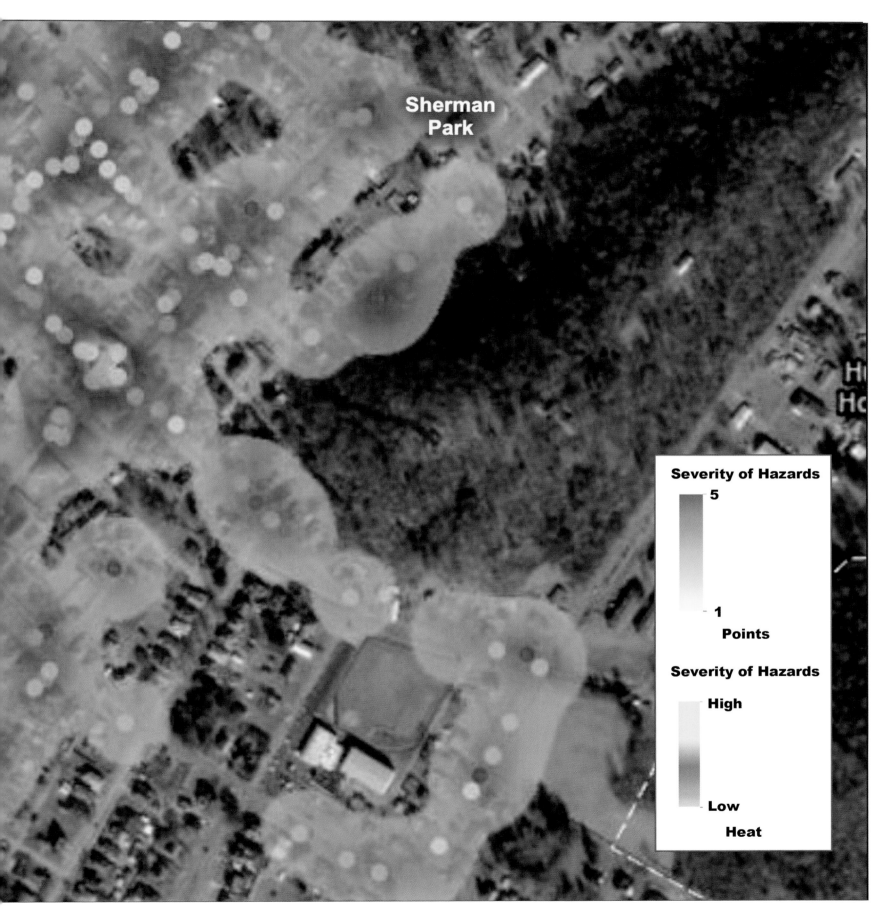

Sherman
Park

H
Ho

Severity of Hazards

5

1

Points

Severity of Hazards

High

Low

Heat

OKALOOSA COUNTY COMMISSIONERS DISTRICTS

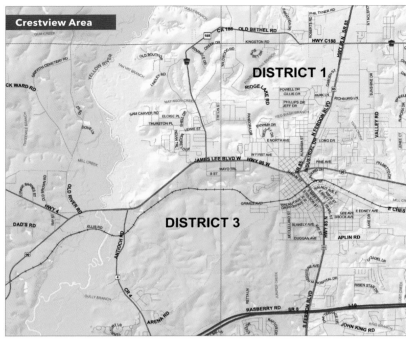

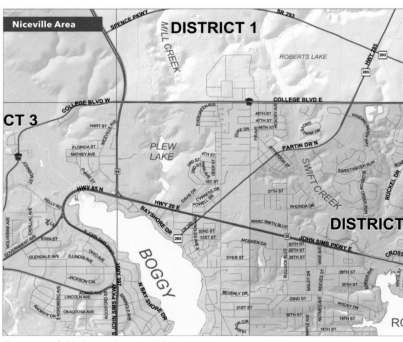

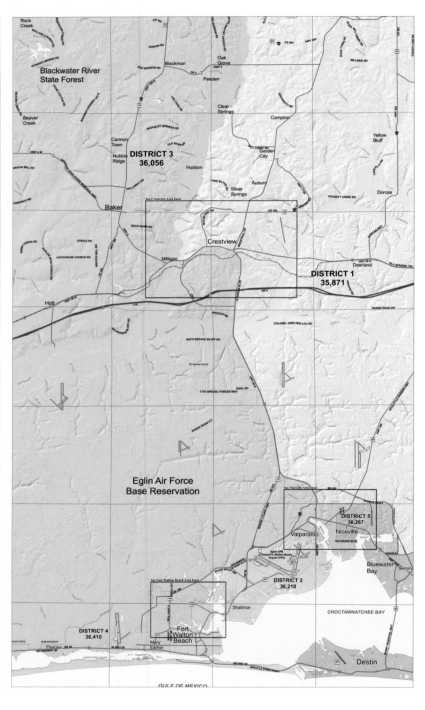

Courtesy of Okaloosa County GIS.

Okaloosa County
Shalimar, Florida, USA
By Colleen A. Pardue

This Okaloosa County Commissioners District map displays the current commissioners district areas as they were determined in the previous census. The map also shows streets, water bodies, points of interest, and township range lines. It contains three insets, showing more densely populated areas, which have more than one district boundary. The most recent census (2010) data tables show relevant demographic data at the bottom of the map.

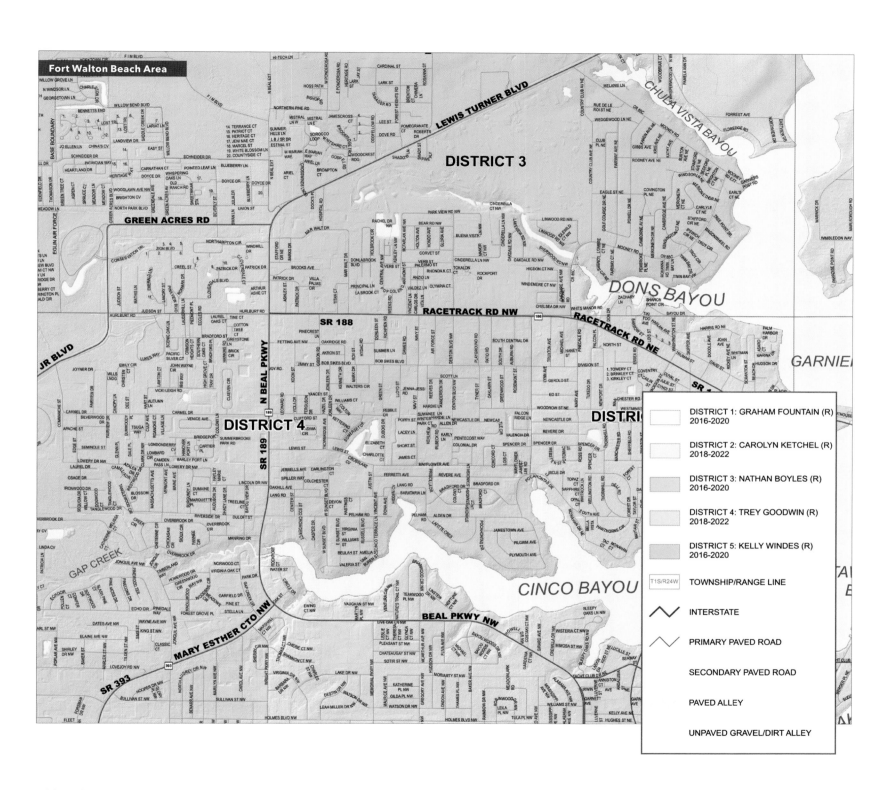

Fort Walton Beach Area

DISTRICT 3

DISTRICT 4

DONS BAYOU

CINCO BAYOU

GARNIER

	DISTRICT 1: GRAHAM FOUNTAIN (R) 2016-2020
	DISTRICT 2: CAROLYN KETCHEL (R) 2018-2022
	DISTRICT 3: NATHAN BOYLES (R) 2016-2020
	DISTRICT 4: TREY GOODWIN (R) 2018-2022
	DISTRICT 5: KELLY WINDES (R) 2016-2020
T1S/R24W	TOWNSHIP/RANGE LINE
	INTERSTATE
	PRIMARY PAVED ROAD
	SECONDARY PAVED ROAD
	PAVED ALLEY
	UNPAVED GRAVEL/DIRT ALLEY

CONTACT
Eddie Quinlan
equinlan@myokaloosa.com

SOFTWARE
ArcGIS Desktop 10.7

DATA SOURCES
Okaloosa County GIS road centerlines, composite land records, commissioner districts, and water bodies

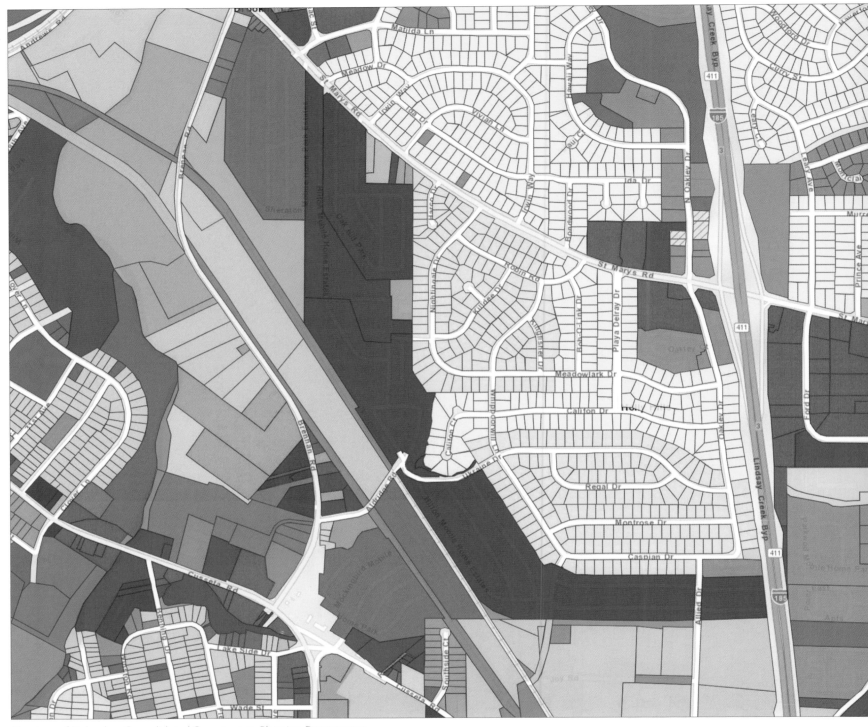

Courtesy of Columbus Consolidated Government, Planning Department.

PUBLIC-FACING ZONING APPLICATION FOR MUSCOGEE COUNTY

Columbus Consolidated Government,
Planning Department
Columbus, Georgia, USA
By Dave Cooper

Designed to be a "one-stop shop" of sorts, the Public-Facing Zoning Application for Muscogee County provides the zone codes for parcels in Columbus, Georgia. This application also provides a rich source of data, such as future and existing land use and flood zones, which allows users to research their area of interest in detail.

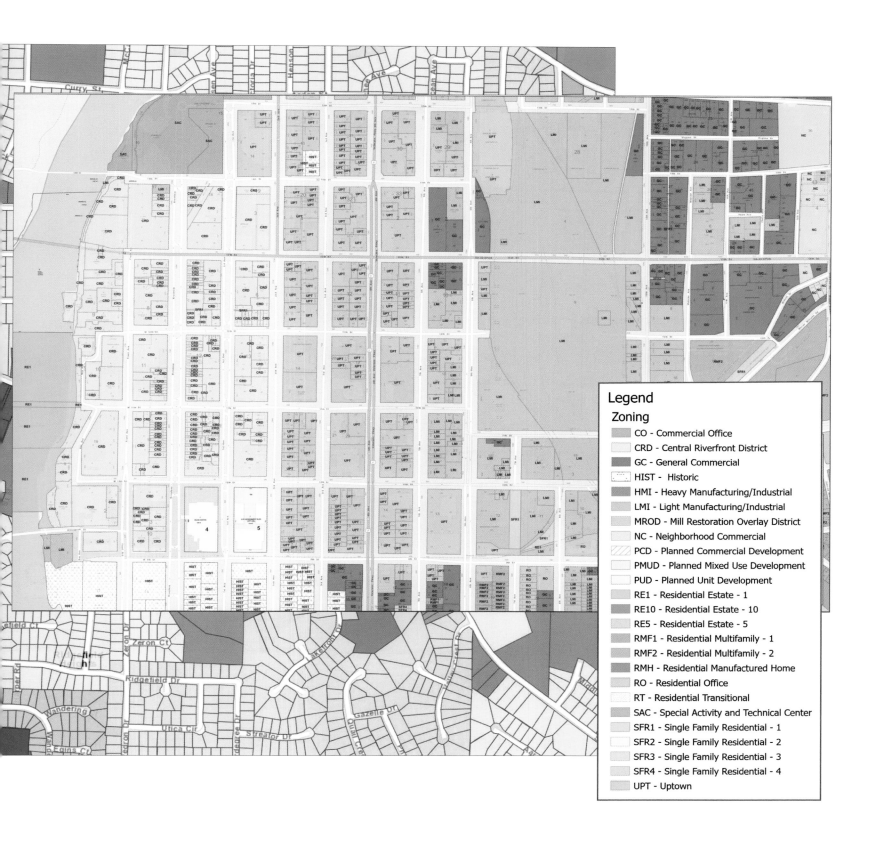

Legend

Zoning

- CO - Commercial Office
- CRD - Central Riverfront District
- GC - General Commercial
- HIST - Historic
- HMI - Heavy Manufacturing/Industrial
- LMI - Light Manufacturing/Industrial
- MROD - Mill Restoration Overlay District
- NC - Neighborhood Commercial
- PCD - Planned Commercial Development
- PMUD - Planned Mixed Use Development
- PUD - Planned Unit Development
- RE1 - Residential Estate - 1
- RE10 - Residential Estate - 10
- RE5 - Residential Estate - 5
- RMF1 - Residential Multifamily - 1
- RMF2 - Residential Multifamily - 2
- RMH - Residential Manufactured Home
- RO - Residential Office
- RT - Residential Transitional
- SAC - Special Activity and Technical Center
- SFR1 - Single Family Residential - 1
- SFR2 - Single Family Residential - 2
- SFR3 - Single Family Residential - 3
- SFR4 - Single Family Residential - 4
- UPT - Uptown

CONTACT
Dave Cooper
davidcooper@columbusga.org

SOFTWARE
ArcGIS Desktop 10.5,
ArcGIS Online

DATA SOURCES
Columbus Consolidated
Government

HIGH-ACCURACY GEOREFERENCED TAX MAP WITH PARCEL FABRIC

Colliers Engineering & Design, Inc., Red Bank, New Jersey, USA, By Amanda Paton and Kristin Esposito

Colliers Engineering & Design uses the power of geographic information systems (GIS) to improve tax mapping and parcel data. Our professionals took a hand-drafted tax map from the 1930s and transformed it into a GIS parcel fabric dataset that is spatially accurate to within a few feet.

New Jersey tax maps have been historically hand drafted or created with computer-aided design (CAD) software, but the City of Elizabeth's tax map sheets were created with GIS. GIS retains the accuracy and practicality of CAD tax maps while combining the power of a database that enables each parcel to have its own specific

attributes recorded directly in the data. The parcels belong to the parcel fabric dataset, which offers a way to maintain a seamless surface of connected parcels that honor survey-grade control points. The parcel fabric was drafted based on the legal description of each parcel and overlaid on high-accuracy aerial imagery

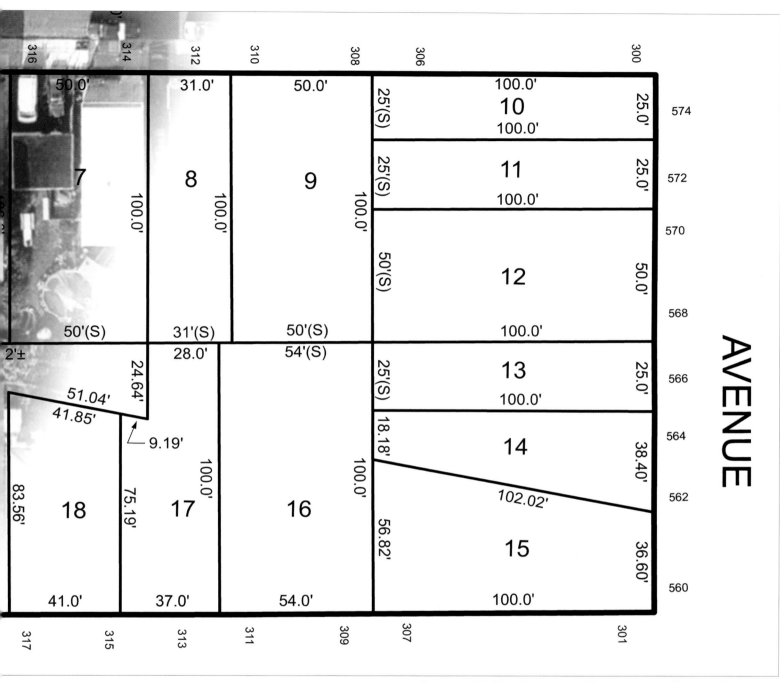

Courtesy of Colliers Engineering & Design, Inc. and Elizabeth, New Jersey.

from Nearmap. Having this high-accuracy parcel data in GIS allows us to overlay and analyze the parcels with other data sources, such as flood hazard areas, zoning data, utility data, and more. The GIS parcel data can be joined to the tax assessment database, which will allow the city to easily query owner information, tax history, and additional property record information from the GIS.

With today's landscape changing so quickly, it's crucial to be able to overlay and visualize data spatially, while also having instant access to the attributes behind it.

CONTACT
Amanda Paton
amanda.paton@
colliersengineering.com

SOFTWARE
ArcGIS Desktop 10.6

DATA SOURCES
Colliers Engineering
& Design, Inc.

JAKARTA COVID-19 GROWTH IN A SPACE-TIME CUBE

Esri Indonesia
Jakarta Selatan, DKI Jakarta, USA
By Yoga Abdilah

CONTACT
Yoga Abdilah
yabdilah@esriindonesia.co.id

SOFTWARE
ArcGIS Pro

DATA SOURCES
Pemerintah Provinsi DKI Jakarta

In Indonesia, the greater Jakarta area has the largest number of confirmed cases of COVID-19 and includes the province with the greatest number of new confirmed daily COVID-19 cases. Using a space-time cube, we can see this map shows hot spots and cold spots of confirmed cases using location and time as the parameters.

Courtesy of Lucas Suryana and Eufrasia Diatmiko.

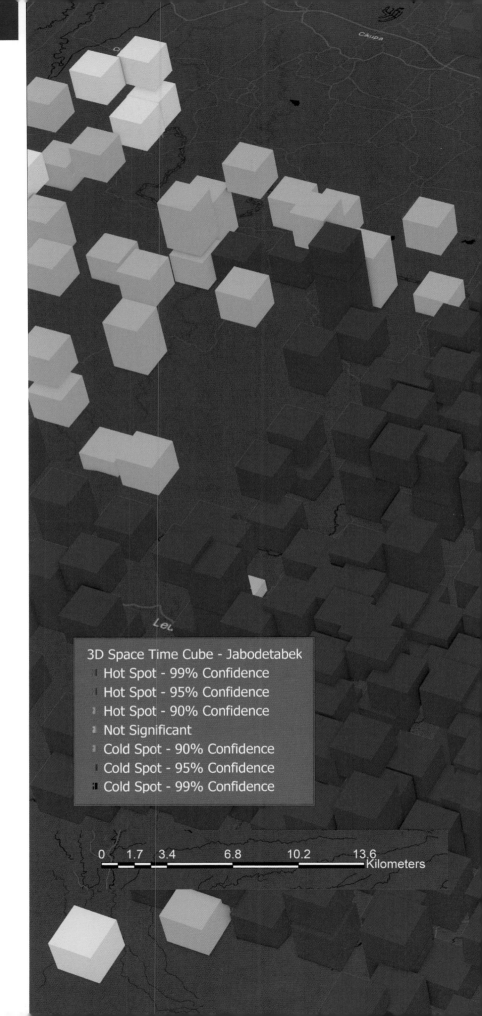

3D Space Time Cube - Jabodetabek
Hot Spot - 99% Confidence
Hot Spot - 95% Confidence
Hot Spot - 90% Confidence
Not Significant
Cold Spot - 90% Confidence
Cold Spot - 95% Confidence
Cold Spot - 99% Confidence

0   1.7   3.4   6.8   10.2   13.6
Kilometers

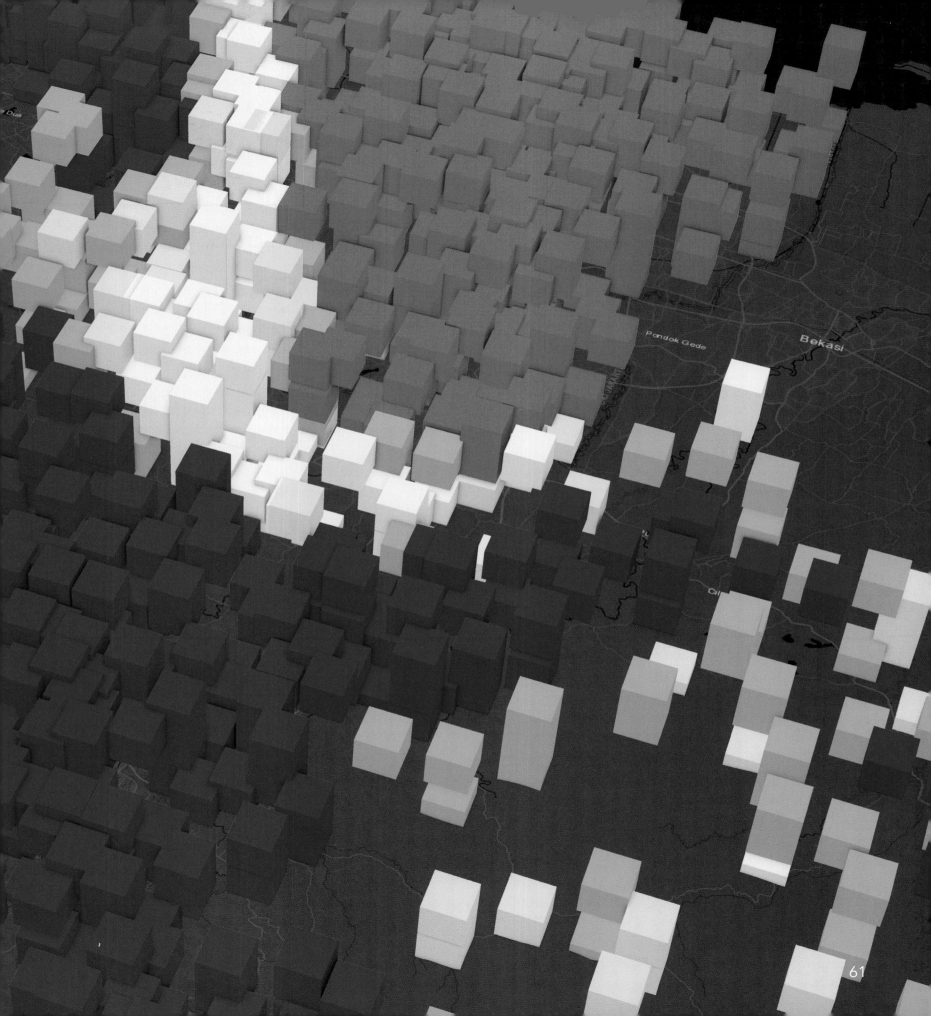

TAMUCC COVID-19 RESPONSE DASHBOARD

Texas A&M University, Corpus Christi (TAMUCC)
Corpus Christi, Texas, USA
By Bryan Gillis

CONTACT
Bryan Gillis
bryan.gillis@tamucc.edu

SOFTWARE
ArcGIS Dashboard

DATA SOURCES
COVID-19 data: Conrad Blucher Institute, Texas Department of State Health Services, Johns Hopkins University

This COVID-19 dashboard was launched in March 2020 to initially provide information to the public concerning public school district closures as a result of COIVD-19 in Texas. As the virus rapidly evolved, so did our dashboard, changing its focus from school closures to the spread of the virus across Texas counties.

With our objective set to inform the public, we started with a much simpler dashboard. Now we are also calculating recovered and active cases for every county in the state of Texas based on World Health Organization formulas, showing trauma service area hospital data, including breakdowns of probable cases and testing efforts. Thanks to Johns Hopkins University, we are also showing comparisons with bordering states and most-affected states.

However, our work is still not done. We recently added new maps and layers to our dashboard that show normalized cases by county population for cumulative, active, recovered, and fatal cases. We also plan to add vaccine data once it becomes available.

Courtesy of Texas A&M University, Corpus Christi.

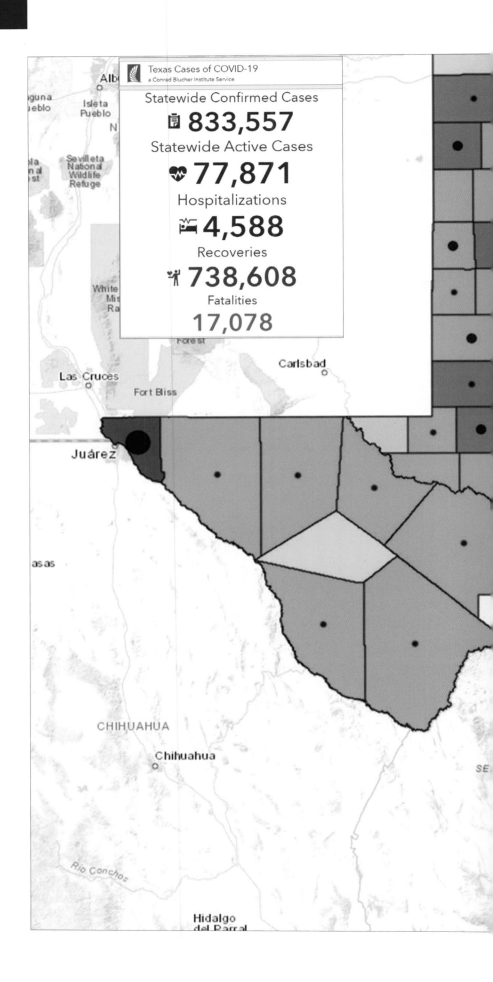

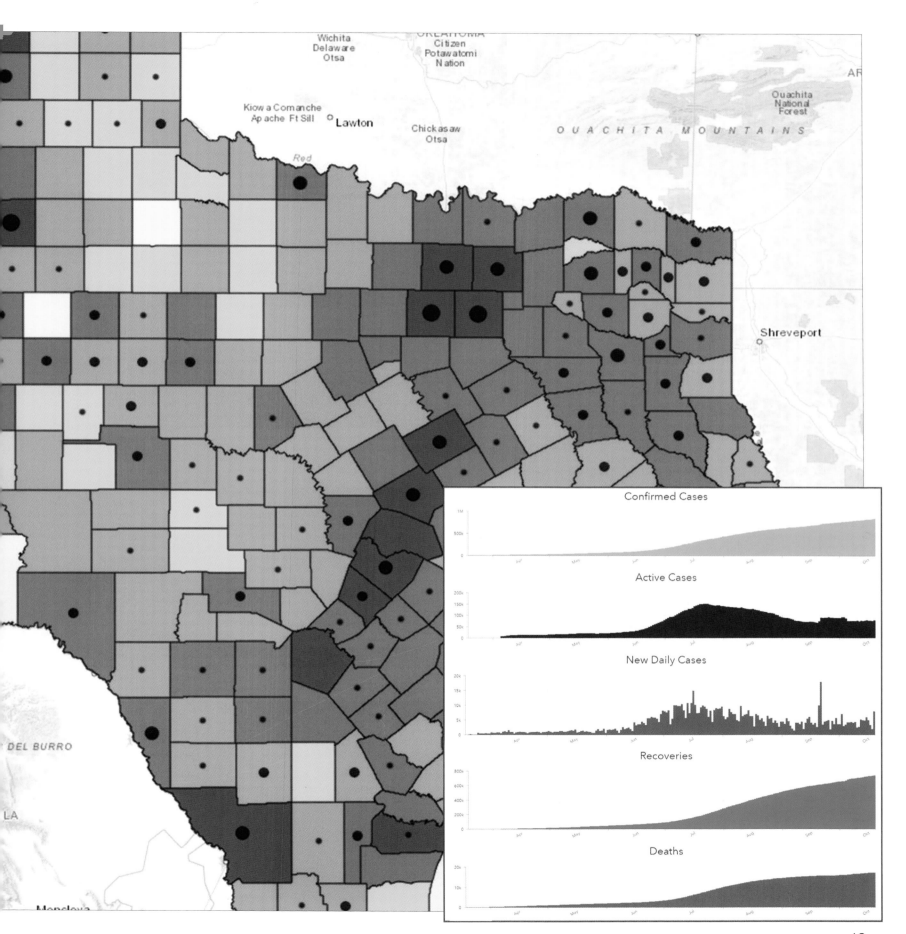

TACOMA COVID-19 RESPONSE TOOL STORYTELLING MAP

City of Tacoma
Tacoma, Washington, USA
By Adriana Abramovich

CONTACT

Adriana Abramovich
aabramovich@cityoftacoma.org

SOFTWARE

ArcGIS Pro

DATA SOURCES

Washington State Health Department, Tacoma-Pierce County Health Department, WHO, CDC, 1point3acres, Worldmeters.info, Johns Hopkins University, National Provider Identifier Registry, and Support Small Business Seattle

The maps and dashboard embedded in this story provide the means to investigate and understand the spread of COVID-19, guide control measures, and assess strategies to control COVID-19 response.

The Confirmed Cases Dashboard shows daily updates of confirmed cases and deaths in Washington State by county and area or city boundary. Medical Services helps emergency medical technicians and the public locate hospitals, medical centers, and physicians using the Health Care Provider National Registry (one of the snapshots at the Esri UC 2020 plenary). The web map also includes the number of ICU licensed and utilized beds at the hospital from Definitive Healthcare USA, an Esri partner. Vulnerable Population is one of the maps shown at the Esri UC plenary that indicates the population's demographic makeup at increased risk (60+ years of age) and uses the CDC Vulnerability Index to help response planners and public officials to dispatch support first. Equity Index highlights the city's ranked disparities in five categories: very low to very high opportunities on accessibility, livability, education, and economics. The index assesses factors limiting opportunities and equity in a community and assists in identifying what measures are needed to remedy these impediments to opportunity. Support Puget Sound Business is a joint effort with the City of Seattle and regional agencies to support restaurant owners in adding themselves to a map with their contact information. The public can search by location for nearby restaurants that are open for takeout, delivery, or third-party delivery. Global Perspective contains the Johns Hopkins University COVID-19 dashboard to present the Tacoma region with a world view of COVID-19 cases.

Medical Services—City of Tacoma

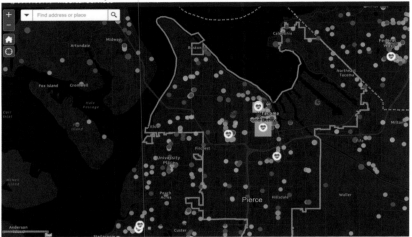

Support Puget Sound-Small Businesses-City of Tacoma, Bureau of Land Management

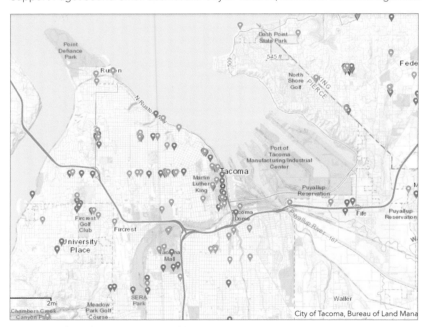

Courtesy of Washington State Health Department, Tacoma-Pierce County Health Department, WHO, CDC, 1point3acres, Worldmeters.info, Johns Hopkins University, National Provider Identifier Registry, and Support Small Business Seattle.

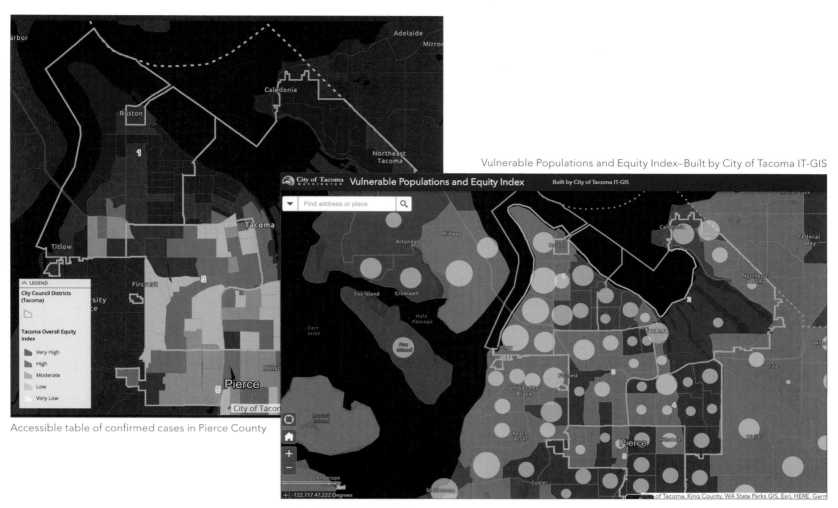

Accessible table of confirmed cases in Pierce County

Vulnerable Populations and Equity Index—Built by City of Tacoma IT-GIS

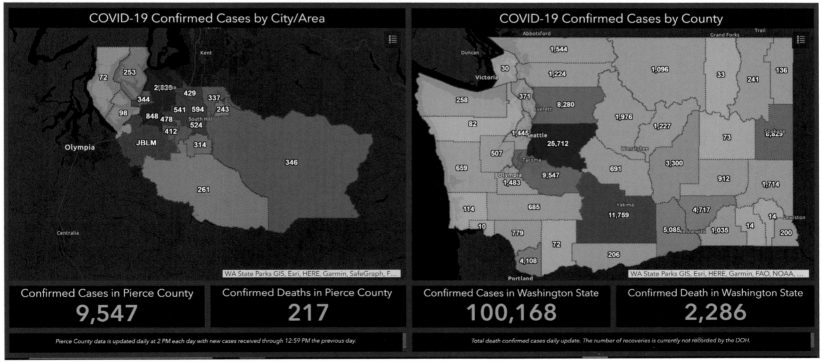

Coronavirus COVID–19 in Tacoma-Pierce County—Built by City of Tacoma IT-GIS

SMART METERS USE IN HOUSEHOLD STATISTICS

UAE Federal Competitiveness and Statistics Authority (FCSA)
Dubai, United Arab Emirates
By Marwa Elkabbany

CONTACT
Marwa Elkabbany
marwa.elkabbany@fcsa.gov.ae

SOFTWARE
ArcGIS Enterprise, ArcGIS Pro

DATA SOURCES
UAE Federal Competitiveness and Statistics Authority (FCSA);
Federal Electricity & Water Authority (FEWA)

The adoption of electricity and water meter data has proven to be an efficient source of household information and land-use distribution. In 2013, the Federal Competitiveness and Statistics Authority (FCSA) completed a project that aimed to define, validate, and acknowledge the methodology of using meter-driven data in identifying household locations and densities. Such digital data adoption methods will minimize field surveys and thus save government resources and time.

This practice represents a smart approach to extract farm and residential unit statistics through deriving data from active utility meters. The identification and classification of active residential/farm water and electricity meters was used to detect the presence of a residential/farm unit. The number of meters was then aggregated at district and subdistrict levels to extract the count of families and farms within the geographical area and deduce family and farm density per square kilometer. The aggregate results for Ajman emirate were compared with Ajman 2017 census results for households, with an accuracy of 96.7 percent.

This approach was used for 2013 and 2017 datasets to present growth occurring after four years.

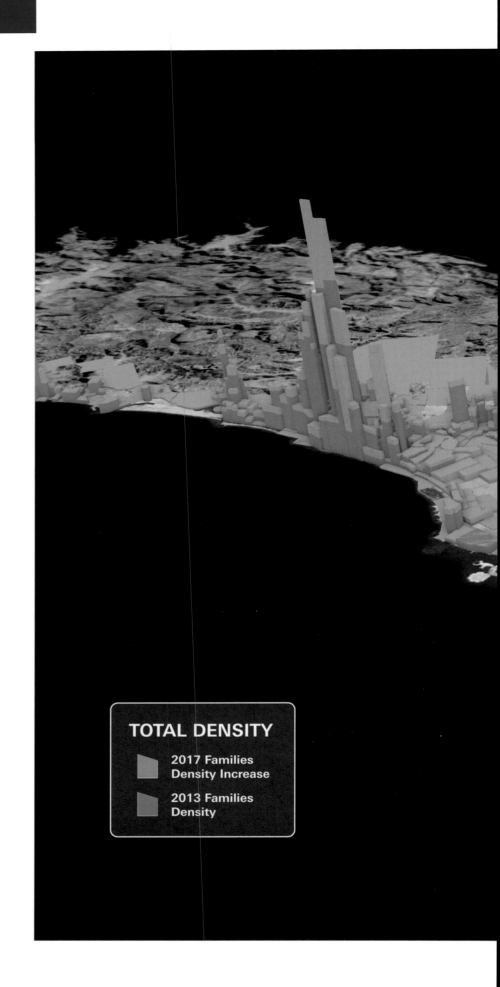

TOTAL DENSITY

2017 Families
Density Increase

2013 Families
Density

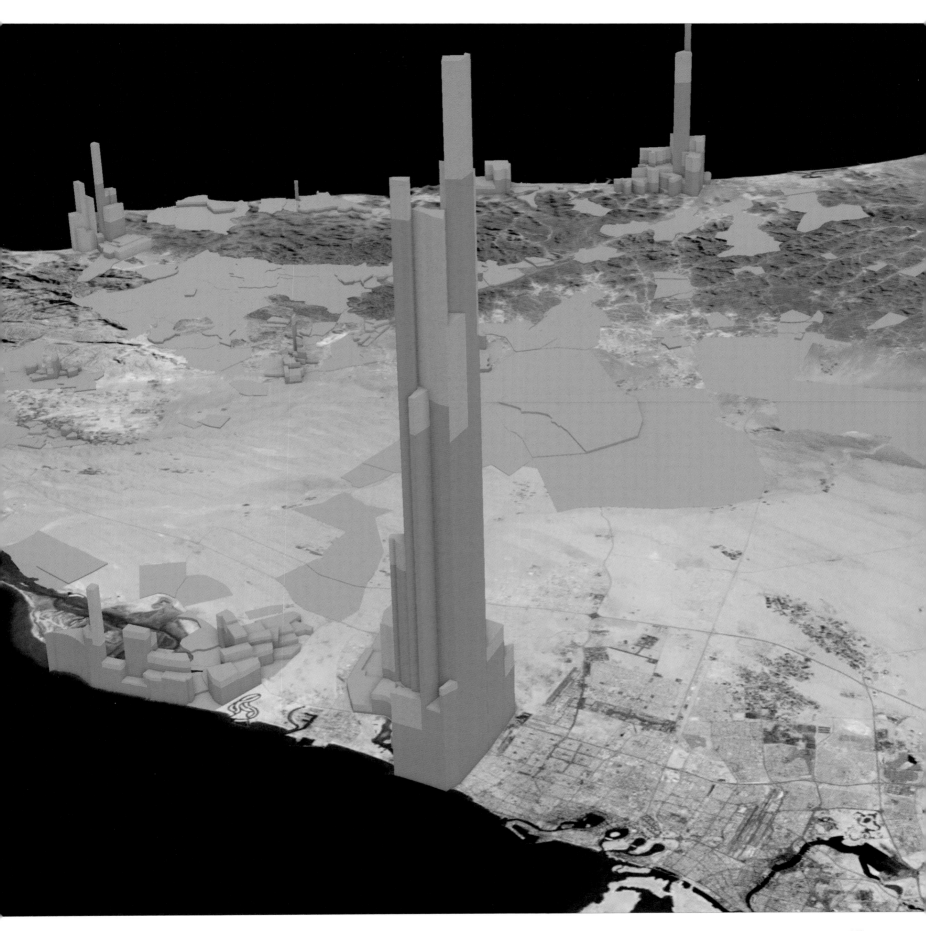

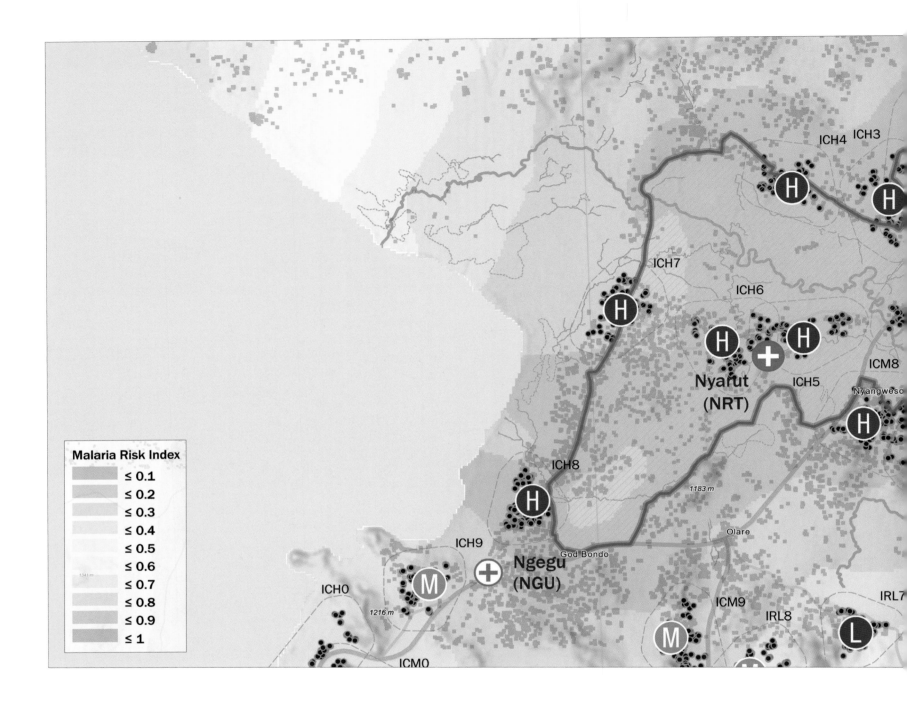

Malaria Risk Index

	≤ 0.1
	≤ 0.2
	≤ 0.3
	≤ 0.4
	≤ 0.5
	≤ 0.6
	≤ 0.7
	≤ 0.8
	≤ 0.9
	≤ 1

STUDY DESIGN OF MALARIA RISK AT HOMA BAY, KENYA

University of California, Irvine
Irvine, California, USA
By Ming-Chieh Lee

This study is based on the potential malaria risk for the Sub-Saharan Africa International Center of Excellence for Malaria Research (ICEMR) project at Homa Bay, Kenya. More than 30 open cohort study clusters in high, moderate, and low categories of transmission risk were randomly selected based on the spatial clustering of house location and predefined criteria. All the houses within the 400 square kilometers of the peninsula have been digitized from high-resolution satellite images.

The irrigation schemes across the Homa Bay Peninsula have also been digitized based on the Kimira-Oluch Small Holder Farm Improvement Project (KOSFIP). The irrigation scheme and its vicinity have been selected as the focus area for monitoring malaria transmission, and the nearby Kandiege subdistrict hospital and 10 other public health facilities were also selected to dedicate the malaria case surveillance.

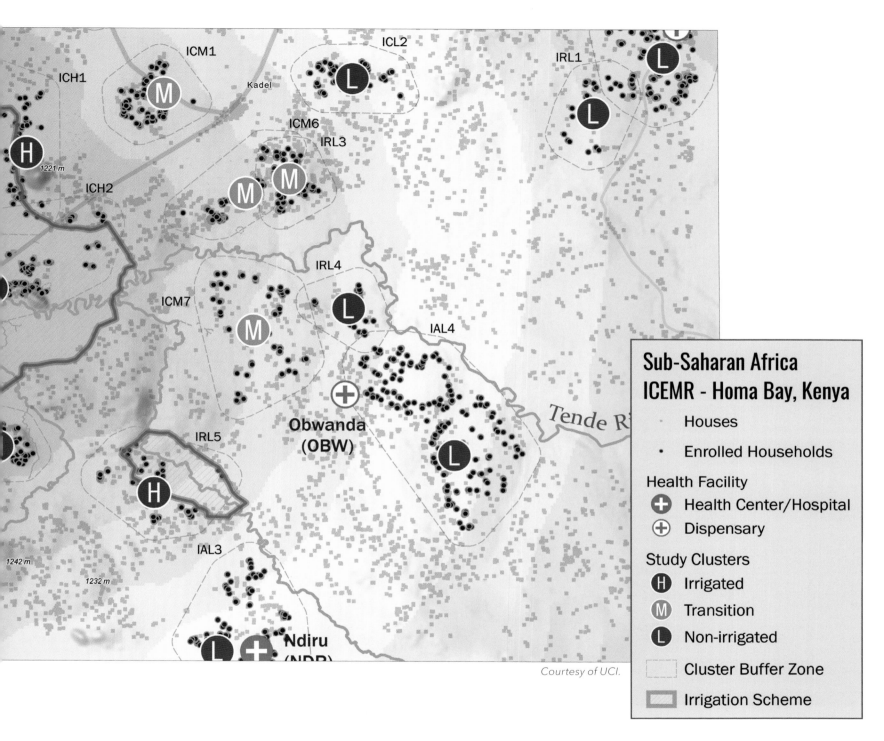

Sub-Saharan Africa ICEMR - Homa Bay, Kenya

- · Houses
- • Enrolled Households

Health Facility
- ⊕ Health Center/Hospital
- ⊕ Dispensary

Study Clusters
- Ⓗ Irrigated
- Ⓜ Transition
- Ⓛ Non-irrigated
- ⬚ Cluster Buffer Zone
- ▨ Irrigation Scheme

Courtesy of UCI.

Based on the distance of each house location to the irrigation scheme, malaria vector abundance, and seasonal normalized difference vegetation index (NDVI), the potential malaria transmission risks were estimated with ArcGIS Pro. The clusters were developed with the radii varying from 0.25 to 1.0 km and the population ranging from 50 to 250 residents and then were marked in red, orange, and green to represent transmission risk categories in high, moderate, and low as well as the irrigated, transition, and nonirrigated zone.

CONTACT
Ming-Chieh Lee
mclee523@gmail.com

SOFTWARE
ArcGIS Pro 2.6

DATA SOURCES
Sub-Saharan Africa ICEMR

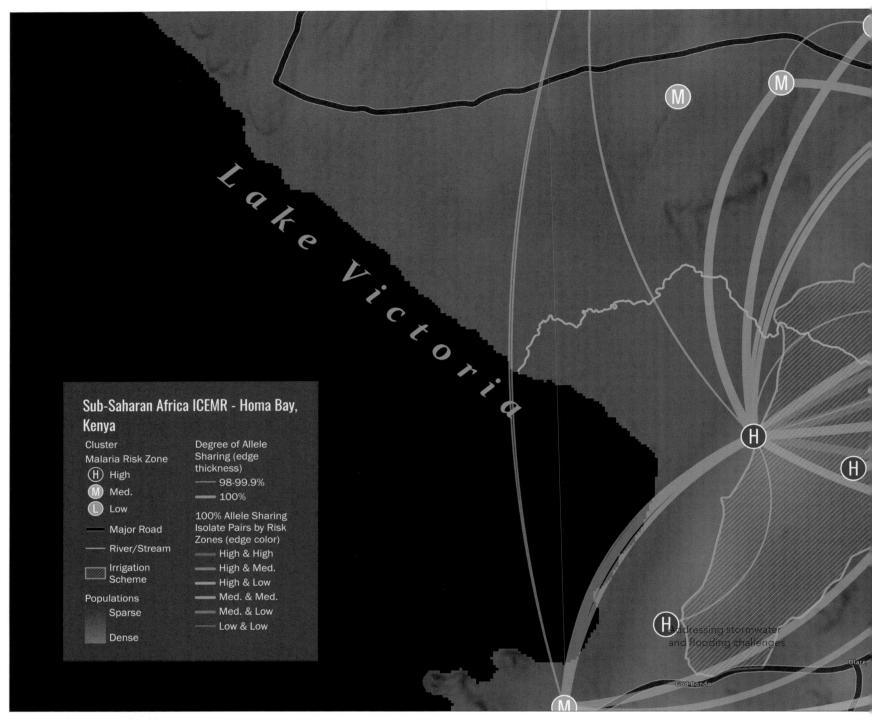

The map legend reads:

Sub-Saharan Africa ICEMR - Homa Bay, Kenya

Cluster
Malaria Risk Zone
- (H) High
- (M) Med.
- (L) Low

— Major Road
— River/Stream
Irrigation Scheme

Populations
Sparse
Dense

Degree of Allele Sharing (edge thickness)
— 98-99.9%
— 100%

100% Allele Sharing Isolate Pairs by Risk Zones (edge color)
— High & High
— High & Med.
— High & Low
— Med. & Med.
— Med. & Low
— Low & Low

Courtesy of University of California, Irvine.

MICROGEOGRAPHIC EPIDEMIOLOGY OF MALARIA PARASITES

University of California, Irvine
Irvine, California, USA
By Ming-Chieh Lee and Elizabeth Hemming-Schroeder

This map illustrates microgeographic epidemiology of malaria parasites in an irrigated area of western Kenya by deep amplicon sequencing. Allele sharing is shown among parasite isolates by cpmp amplicon sequencing of malaria parasites in the Oluch irrigation scheme, Homa Bay, Kenya. The highly related isolates are shown by geographic locality. Dots indicate individual parasite isolates. Lines connect isolates that share >98 percent of polymorphic alleles.

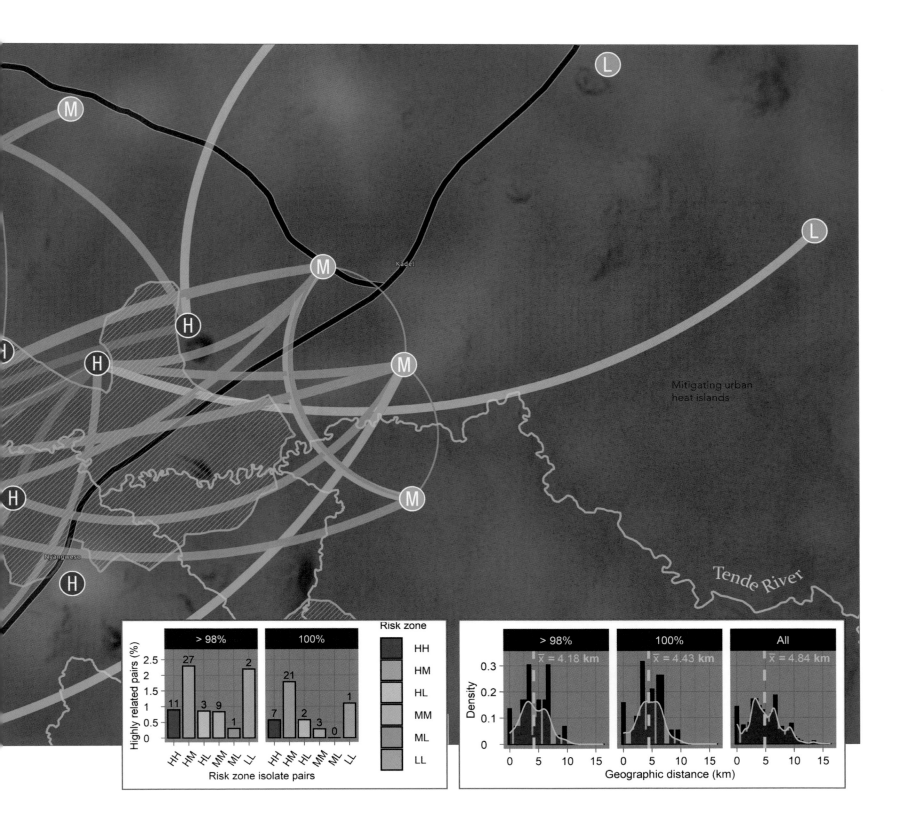

Thin lines indicate 98 to 99.9 percent allele sharing; thick lines indicate 100 percent allele sharing. Allele sharing is highest between high and medium risk zones. Geographic distance does not explain the pattern of allele sharing.

CONTACT
Ming-Chieh Lee
mingchil@uci.edu

SOFTWARE
ArcGIS Pro 2.6

DATA SOURCES
Sub-Saharan Africa ICEMR

MIGRATION CRISIS IN COLOMBIA

The Fletcher School of Law and Diplomacy, Cambridge, Massachusettes, USA
By Jake Mendales

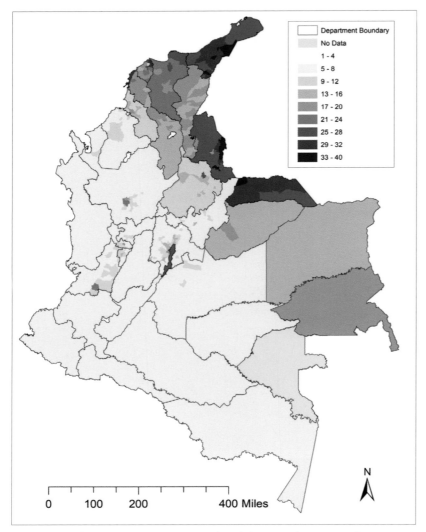

Distribution of Venezuelan Populations Vulnerability Score by Municipality

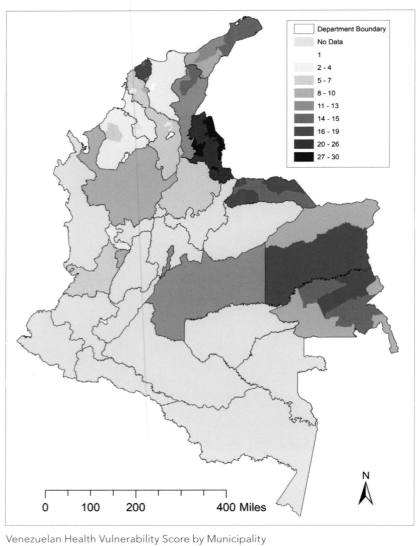

Venezuelan Health Vulnerability Score by Municipality

Venezuela is suffering from one of the Western Hemisphere's worst humanitarian crises in recent memory. More than 10 percent of the country's population (over 3 million people) have fled into neighboring countries in search of food, security, and stability. As the political and economic situation continues to deteriorate, an increasingly authoritarian regime shows more interest in maintaining power than helping its citizens. Millions more are likely to flee, and soon Venezuela's migrant crisis could be on a par with Syria. This humanitarian crisis is becoming a regional crisis, and more countries must shoulder the responsibility of caring for, treating, and protecting Venezuelan refugees. This is coming not just at a financial cost but, increasingly, a political and social cost.

This project aims to evaluate and assess the capacity and risks for Venezuela's neighbor to the west, Colombia, which continues to absorb more and more fleeing migrants. More than 3 million Venezuelans have arrived in Colombia through regular and irregular means, though this project relies on data from Colombia's Administrative Registry of Venezuelan Migrants, which reports 442,462 registered individuals

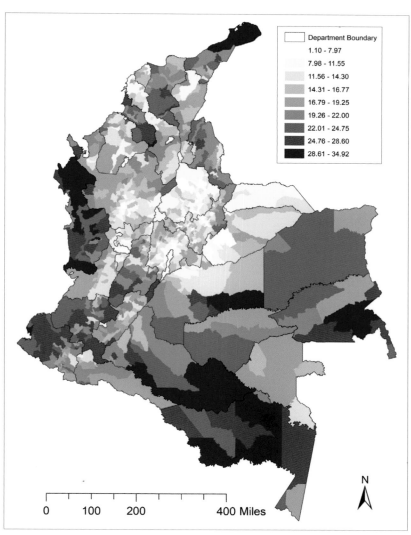

Colombian Socio-Economic and Infrastructure Vulnerability Score by Municipality

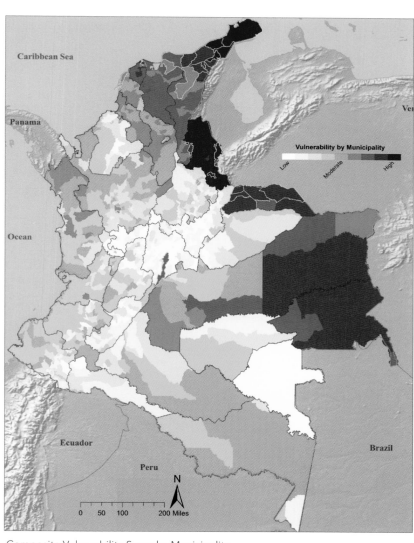

Composite Vulnerability Score by Municipality

as of July 2018. By identifying the areas of Colombia that possess a higher vulnerability for its own citizens, we can overlay information about where Venezuelan migrants are relocating and examine where there will probably be greater strain on municipal governments and social services. Such spatial analysis tools can be used to pinpoint areas for interventions, aid, and development projects.

Courtesy of The Fletcher School of Law and Diplomacy.

CONTACT
Jake Mendales
jmendales@gmail.com

SOFTWARE
ArcGIS Desktop

DATA SOURCES
DANE, Migración Colombia,
MINSALUD,
Government of Colombia,
Esri, Open Street Map

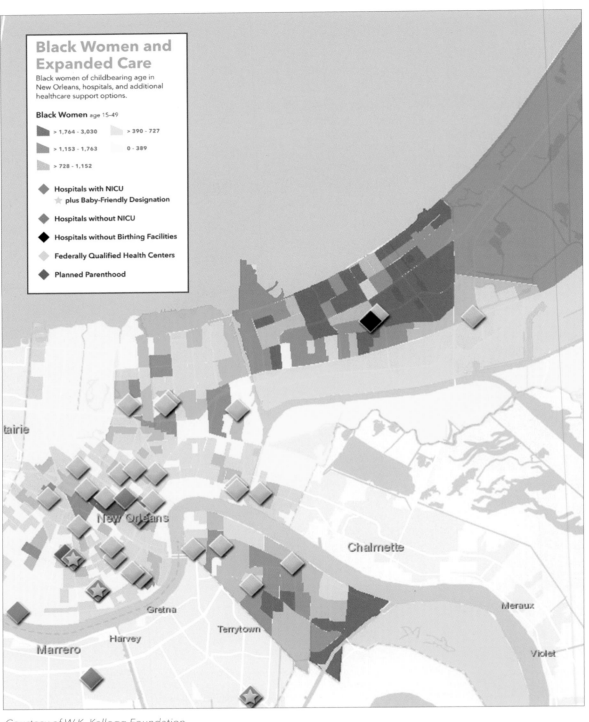

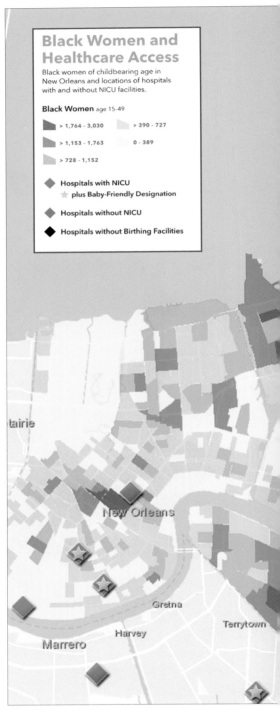

Courtesy of W.K. Kellogg Foundation.

HEALTHCARE ACCESS, RACIAL EQUITY, AND MATERNAL MORTALITY

W.K. Kellogg Foundation
Battle Creek, Michigan, USA
By Luis Velez, Amanda Otter,
and Dylan Peterson

In Louisiana, maternal mortality for African American mothers is four times as high as the ratio for non-Hispanic white mothers. This map shows where women are having babies in New Orleans, Louisiana, where African American women live, and where the healthcare facilities are that provide birthing facilities and have neonatal intensive care units.

Births and Healthcare Access

Births in the last 12 months for the total population in New Orleans, Black women of childbearing age, and access to hospitals.

Fertility
Number of women who have given birth in the last 12 months.

> 222   170   100   50   < 0

Black Women age 15-49

- > 1,764 - 3,030
- > 1,153 - 1,763
- > 728 - 1,152
- > 390 - 727
- 0 - 389

- ◆ Hospitals with NICU
  - ★ plus Baby-Friendly Designation
- ◆ Hospitals without NICU
- ◆ Hospitals without Birthing Facilities

CONTACT
Ross Comstock
ross.comstock@wkkf.org

SOFTWARE
Map Viewer, ArcGIS
StoryMaps

DATA SOURCES
American Community Survey

TEXAS WORKFORCE COMMISSION UI CLAIMANT DASHBOARD

Texas Workforce Commission
Austin, Texas, USA
By Donovan Chudej, Bradley Spears, and Todd Mills

CONTACT
Mark Duksta
mark.duksta@twc.state.tx.us

SOFTWARE
ArcGIS Pro 2.6, ArcGIS Online

DATA SOURCES
Texas Workforce Commission, Bureau of Labor Statistics,
Texas Legislative Council, US Census Bureau,
Texas State Demographer

In response to the COVID-19 pandemic, the Texas Workforce Commission created an interactive online mapping application that displays Unemployment Insurance (UI) claim information by geographic region.

UI claim counts can be displayed by workforce development area, county, zip code, Texas House and Senate districts, and US Congressional districts. Users navigate the application by clicking the appropriate tab at the top of the page for the region to investigate.

On each page, users will find an interactive, searchable map that displays information with a click on a region. Charts and lists on each page convey information at a glance. Charts and lists are also interactive.

Each map, chart, or list click reveals a pop-up that displays population, UI claim count, the top five UI claim industries, claimant gender, a UI claim bar chart, and legislative areas that include the district representative's name.

Industry counts are not displayed for zip code regions because the granularity of the data may allow individual identification of a claimant. All the application data, which is available for download as a CSV file, is updated weekly.

Courtesy of Texas Workforce Commission.

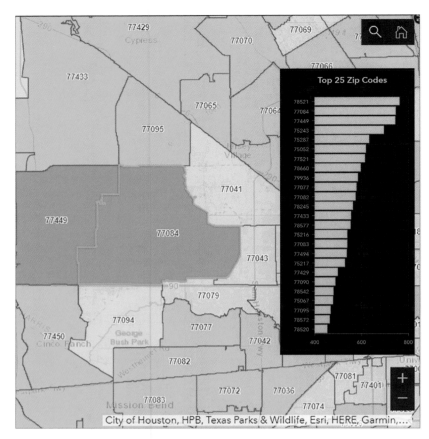

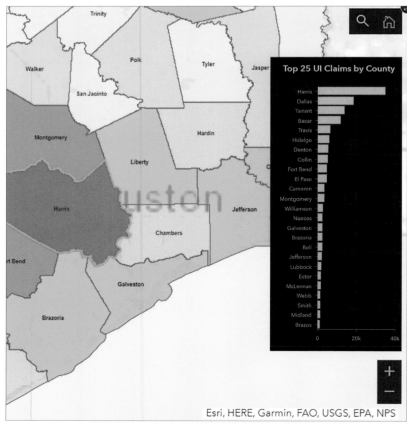

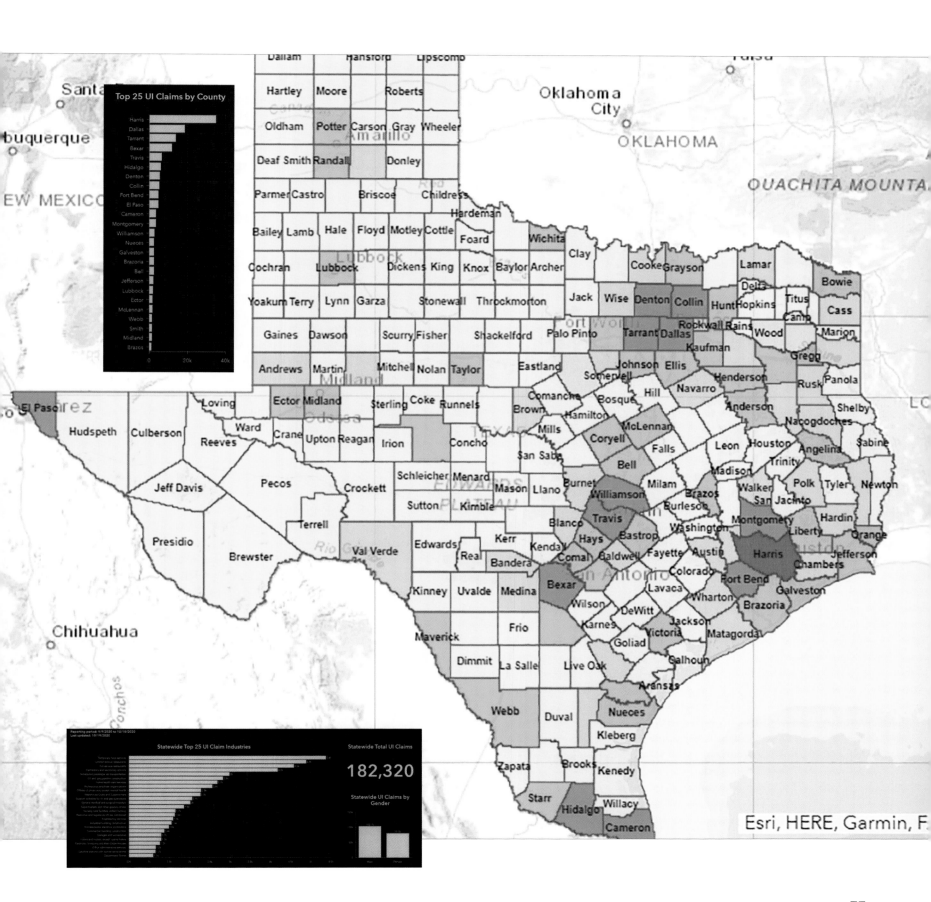

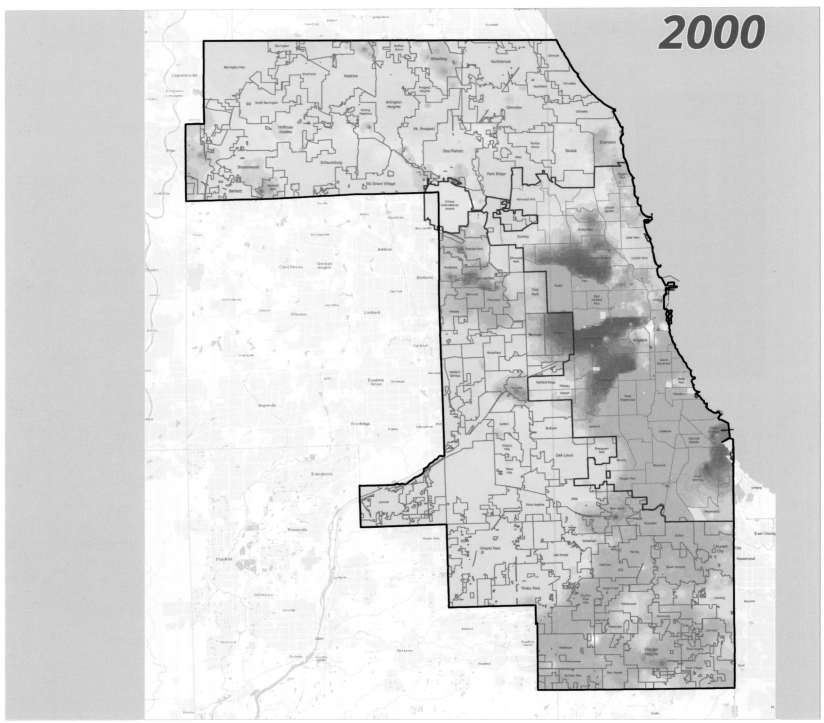

Courtesy of Cook County Government..

THE RACIAL TAPESTRY OF COOK COUNTY

Cook County Government
Chicago, Illinois, USA
By Todd J. Schuble

This map of the racial tapestry of Cook County, Illinois, gives insight into how the density of different demographic groups is distributed throughout the county over time. The three primary groups highlighted (White/Caucasian, Black/African American, and Hispanic/Latino) make up an extremely large portion of the population. The unique colorization technique allows users to see where these different groups merge and blend, in contrast with other areas of the county that are almost completely homogeneous. Certain portions of the county where the colors seem less bold means that a sector of the population that was not displayed

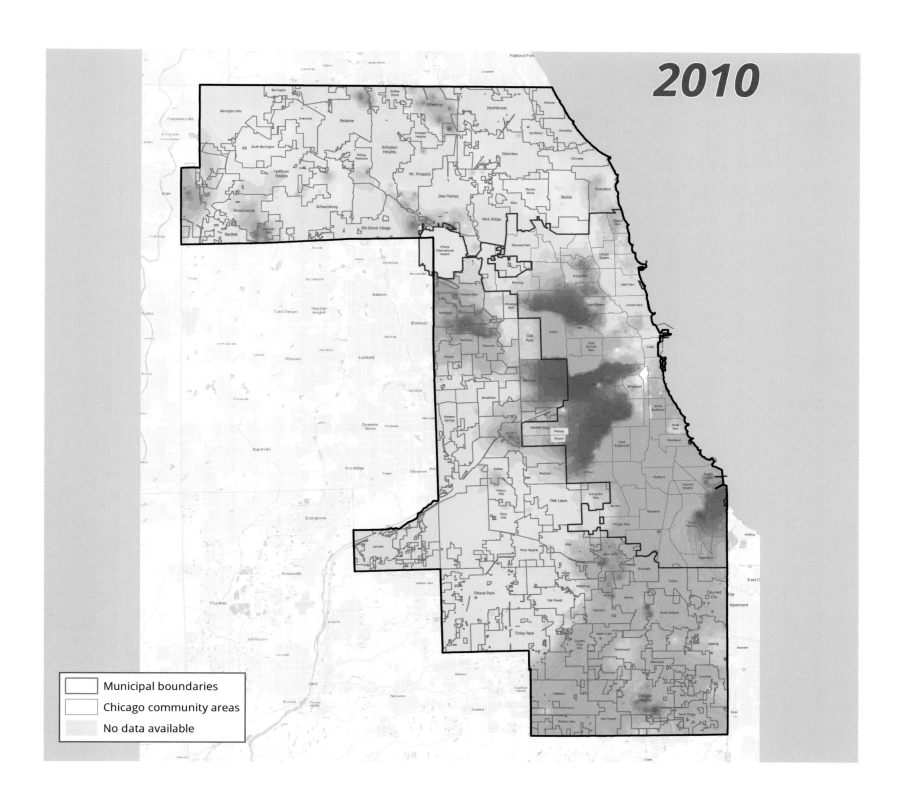

2010

	Municipal boundaries
	Chicago community areas
	No data available

(such as Asian or Native American) lives in those areas. With the 2020 US Census data about to be released, there will be an exciting dynamic to examine how different populations have grown or blended over a 20-year period but also which areas of the county have remained static with regard to their demographic makeup.

CONTACT
Todd J. Schuble
todd.schuble@cookcountyil.gov

SOFTWARE
ArcGIS Desktop

DATA SOURCES
Cook County GIS Department,
US Census Bureau

COMPARING A RANDOM FOREST-BASED PREDICTION OF WINTER WHEAT YIELD TO HISTORICAL YIELD POTENTIAL

Aarhus University, Foulum, Denmark
By Yannik E. Roell, Amélie Beucher, Per G. Møller,
Mette B. Greve, and Mogens H. Greve

CONTACT

Yannik E. Roell
yannik.roell@agro.au.dk

SOFTWARE

ArcGIS Pro 2.4

DATA SOURCES

Danish National Field Trials managed by the Danish
Agricultural Advisory Service (SEGES)

The main map shows winter wheat yield across Denmark. The goal of this map was to update the national yield map. Previously, the national map used only three topsoil properties to predict the yield: clay, silt, and organic carbon. With the new map, 28 environmental variables were used to predict yield across the country with a more sophisticated method. The variables depict landscape information for soil, climate, and terrain, which are all important to consider when growing crops. The new yield map better depicts reality and will be further used to help with land assessments for policy support. The improved and realistic yield prediction can aid in a farmer's decision to buy new land for growing winter wheat.

The collection of smaller maps shows how agriculture production has progressed through time. Denmark has detailed yield potential data for the entire country from as early as the 17th century. The two national land registers in 1688 and 1844 provide valuable information to help determine the natural growth potential before advances in mechanical farm technology were widespread. The updated national yield map was compared with the historical production potential to view how agriculture has changed with space and time. The maps show that advances in technology and farming practices have exceeded historical yield predictions. This indicates that current yield predictions will be unreliable in future years as technology progresses.

Courtesy of Aarhus University and Southern Denmark University.

Random Forest Results

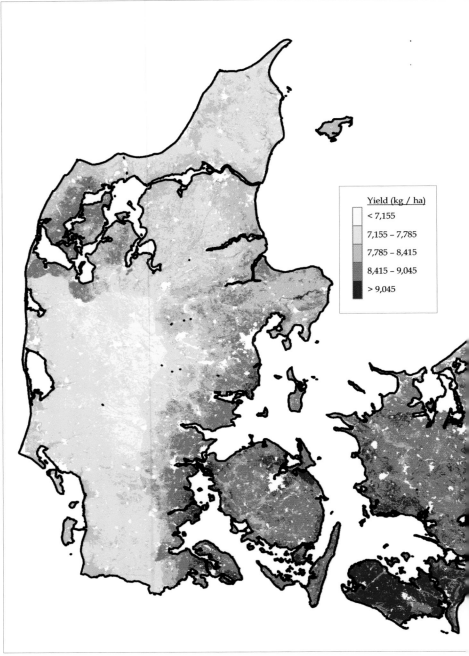

Yield map of winter wheat produced from random forest model

CONTACT

Yannik E. Roell
yannik.roell@agro.au.dk

SOFTWARE

ArcGIS Pro 2.4

DATA SOURCES

Danish National Field Trials managed by the Danish Agricultural Advisory Service (SEGES)

Hard Grain by Parish in 1688

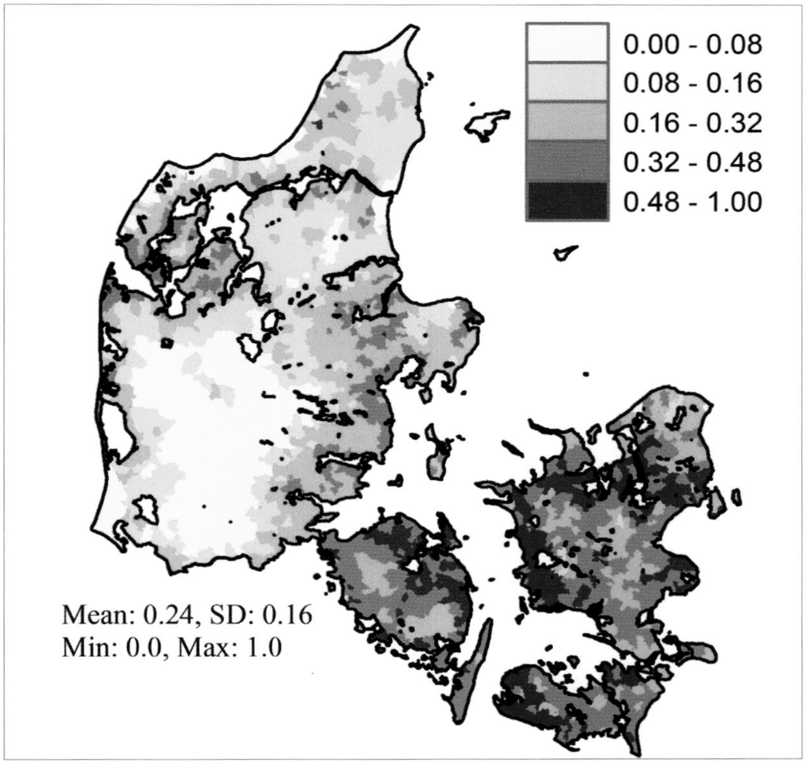
0.00 - 0.08
0.08 - 0.16
0.16 - 0.32
0.32 - 0.48
0.48 - 1.00
Mean: 0.24, SD: 0.16
Min: 0.0, Max: 1.0

Hard Grain by Parish in 1844

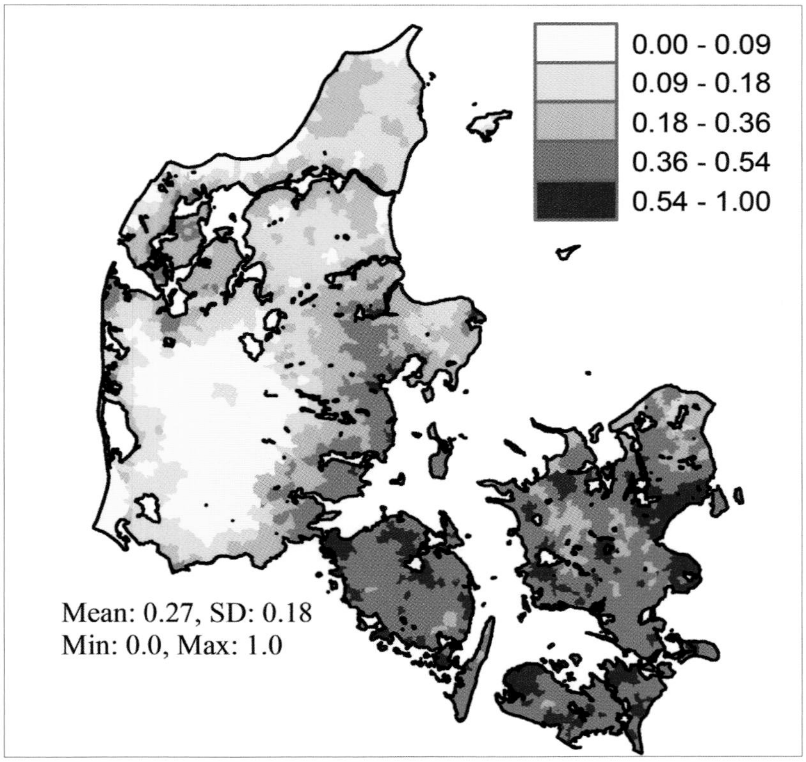
0.00 - 0.09
0.09 - 0.18
0.18 - 0.36
0.36 - 0.54
0.54 - 1.00
Mean: 0.27, SD: 0.18
Min: 0.0, Max: 1.0

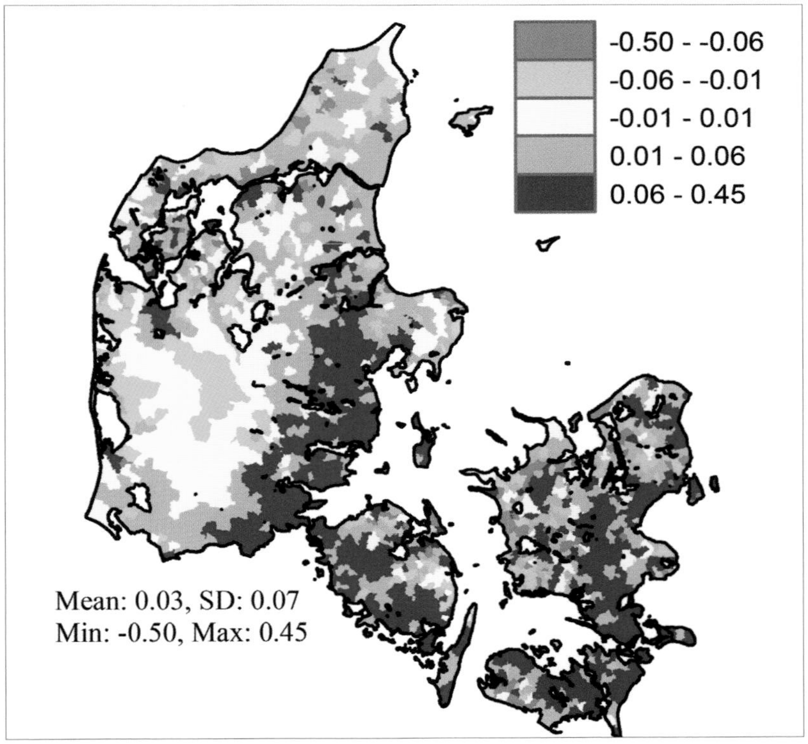
-0.50 - -0.06
-0.06 - -0.01
-0.01 - 0.01
0.01 - 0.06
0.06 - 0.45
Mean: 0.03, SD: 0.07
Min: -0.50, Max: 0.45

Difference between 1844 and 1688

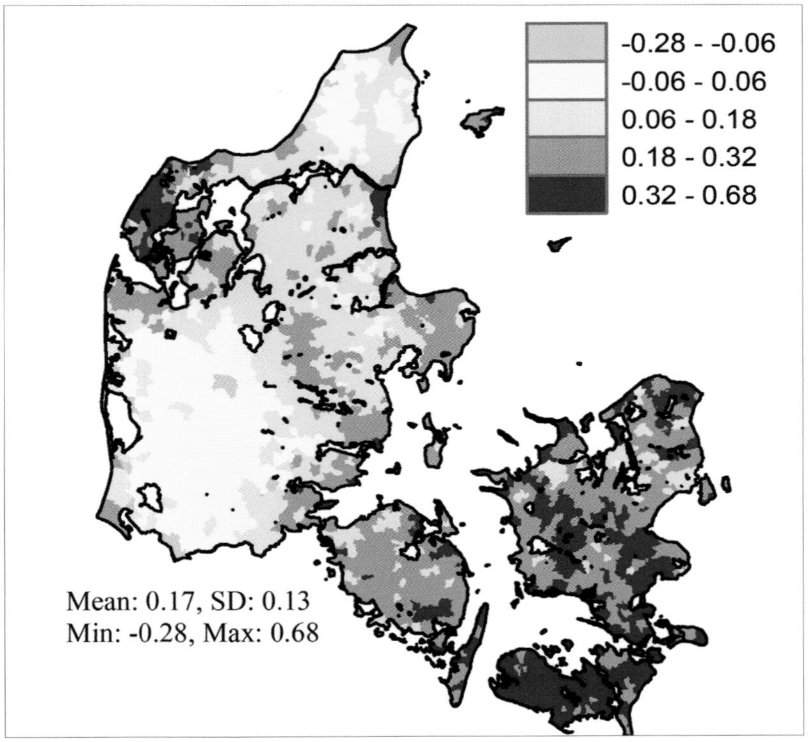
-0.28 - -0.06
-0.06 - 0.06
0.06 - 0.18
0.18 - 0.32
0.32 - 0.68
Mean: 0.17, SD: 0.13
Min: -0.28, Max: 0.68

Difference between Present-day and 1844

MAPPING TREES OUTSIDE FORESTS IN THE CENTRAL UNITED STATES

USDA Forest Service National Agroforestry Center, Lincoln, Nebraska, USA
By Todd Kellerman

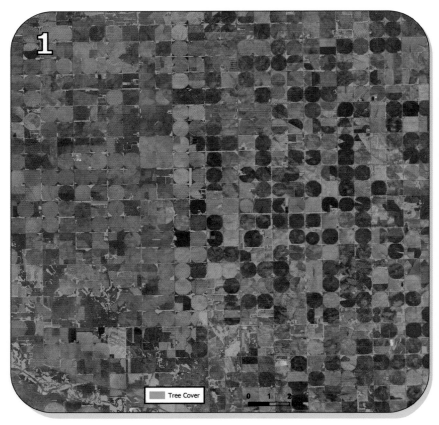

Courtesy of USDA Forest Service National Agroforestry Center.

Trees in the central Unites States are an undercounted yet important ecological resource. Most major natural resource inventories do not account for these trees because of their size and linear nature. Windbreaks and riparian forests serve important environmental, social, and economic resources in this region. The USDA Forest Service National Agroforestry Center in Lincoln, Nebraska, and the USDA Forest Service Forest Inventory & Analysis unit at Northern Research Station in St. Paul, Minnesota, have partnered to develop methods using one-meter National Agriculture Imagery Program (NAIP) imagery to map these trees in this region, including Kansas, Nebraska, South Dakota, and North Dakota. This map highlights an area of eastern Nebraska, with more detailed inset maps to show the prevalence of windbreaks and riparian forested areas. This first-of-its-kind mapping at this scale can serve as a baseline, because little information about these resources is currently available. The information can also help natural resource planners and land managers with conservation efforts going forward.

CONTACT
Todd Kellerman
todd.kellerman@usda.gov

SOFTWARE
ArcGIS Pro 2.4

DATA SOURCES
USDA Forest Service

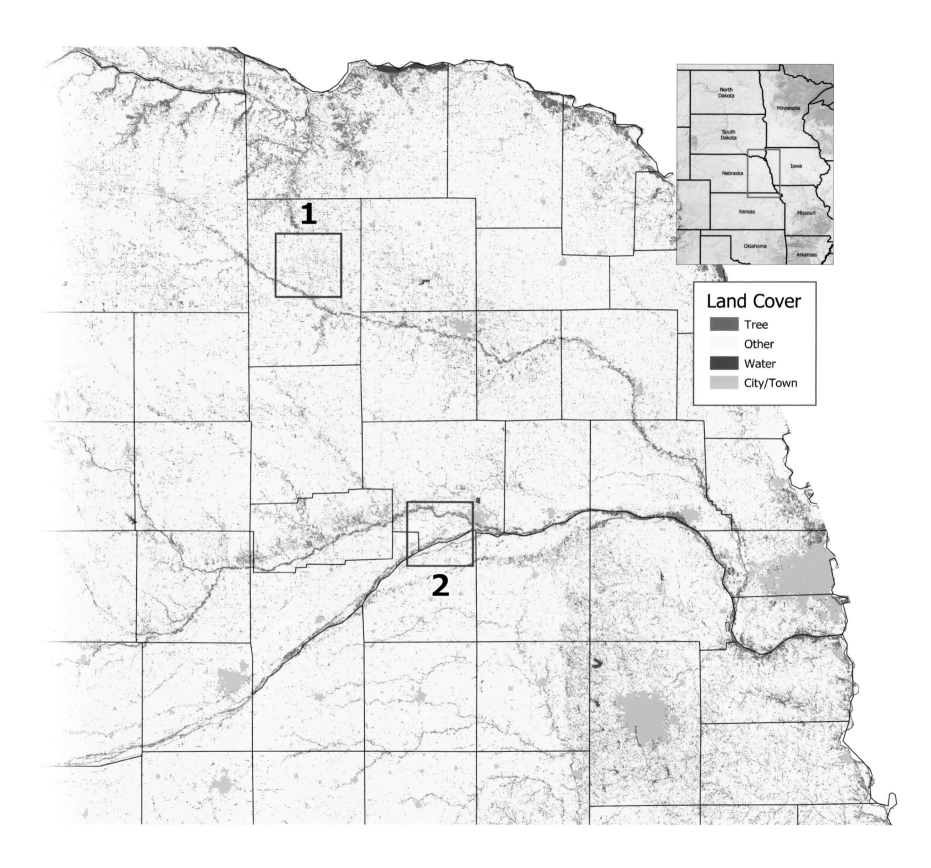

Land Cover

Tree
Other
Water
City/Town

83

COFFEE SUITABILITY IN RWANDA

National Agricultural Export Development Board (NAEB)/Rwanda Kigali, Rwanda
By Alfred Gasore

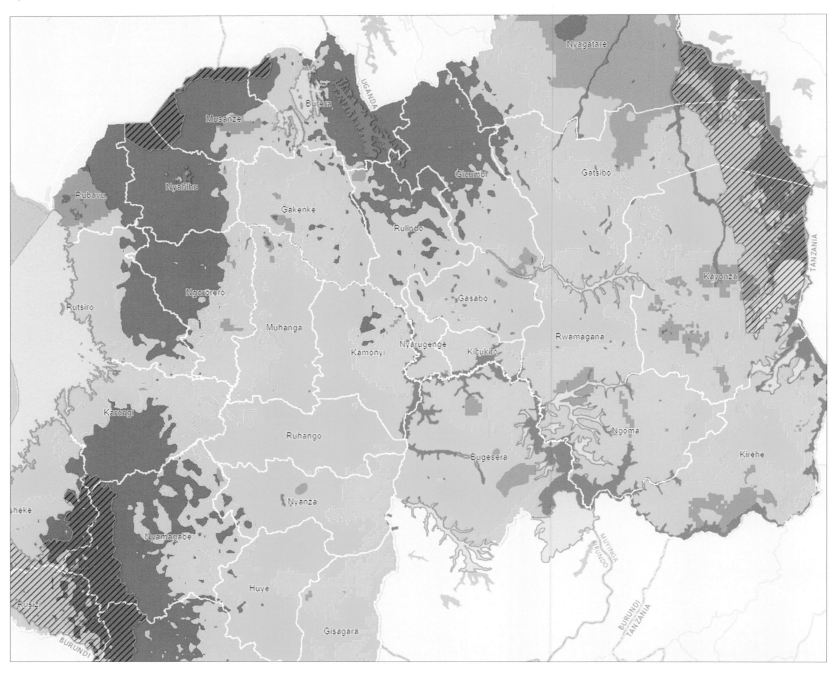

This map compares the suitable areas of coffee plantation in Rwanda for the year 2020, using climatic, topographic, and soil data of Rwanda.

CONTACT
Alfred Gasore
g.alfred@naeb.gov.rw

SOFTWARE
ArcGIS Pro 2.5

DATA SOURCES
NAEB, MINAGRI, Rwanda
Meteorological Agency, RLMUA

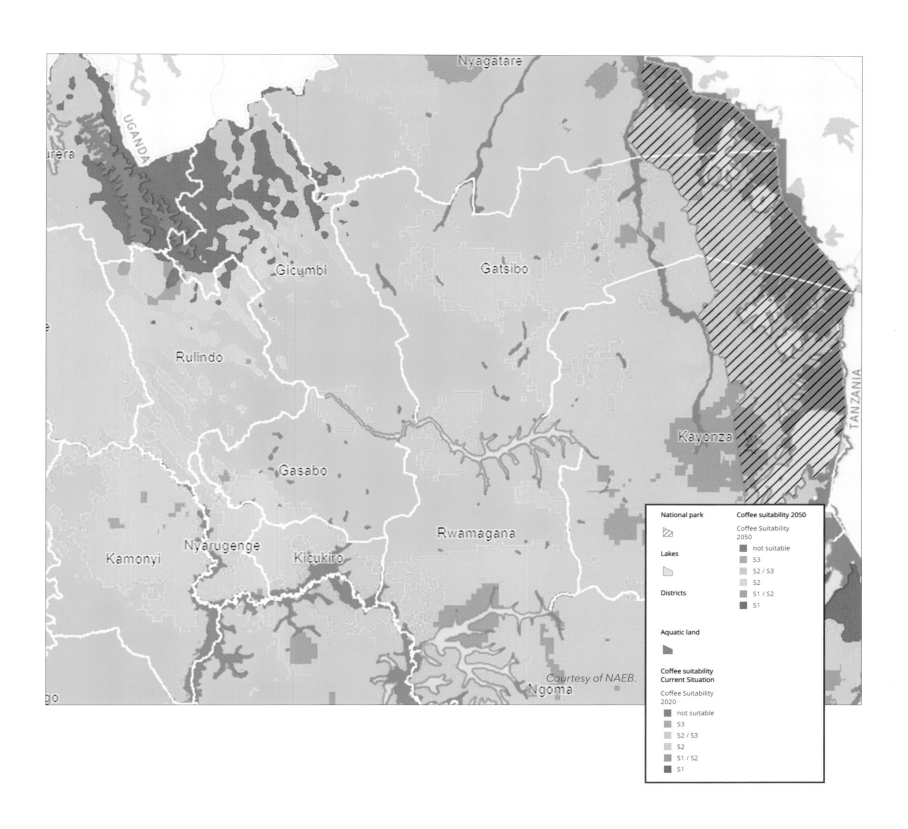

Courtesy of NAEB.

National park

Lakes

Districts

Aquatic land

Coffee suitability
Current Situation

Coffee Suitability
2020

☐ not suitable
☐ S3
☐ S2 / S3
☐ S2
☐ S1 / S2
☐ S1

Coffee suitability 2050

Coffee Suitability
2050

☐ not suitable
☐ S3
☐ S2 / S3
☐ S2
☐ S1 / S2
☐ S1

VINEYARDS OF THE ROGUE VALLEY AMERICAN VITICULTURAL AREA

Jackson County
Medford, Oregon, USA
By Brady R Smith

CONTACT

Brady R Smith
brady.ray.smith@gmail.com

SOFTWARE

ArcGIS Desktop 10.5.1

DATA SOURCES

Jackson County GIS, Oregon DOGAMI

Established in 1991, the Rogue Valley American Viticultural Area (AVA) is the southernmost grape-growing region in Oregon and is contained within the larger Southern Oregon AVA. The Rogue Valley AVA is located within Jackson and Josephine Counties and is approximately 2,100,000 acres in size.

The type of soil can have a profound effect on the quality of wine produced from grapes. The area has mixes of metamorphic, sedimentary, and volcanic soils because three different mountain ranges converge in the Rogue Valley. Since the geology and the resulting soils of the Rogue Valley are diverse, geology is a key consideration in vineyard development for grape growers in the Rogue Valley AVA.

The map shows the location of vineyards, general geology, topography, and the extent of the Rogue Valley AVA. It was created to promote the wine industry in the area and can be used to identify locations where future vineyard development can occur.

Courtesy of Brady Smith.

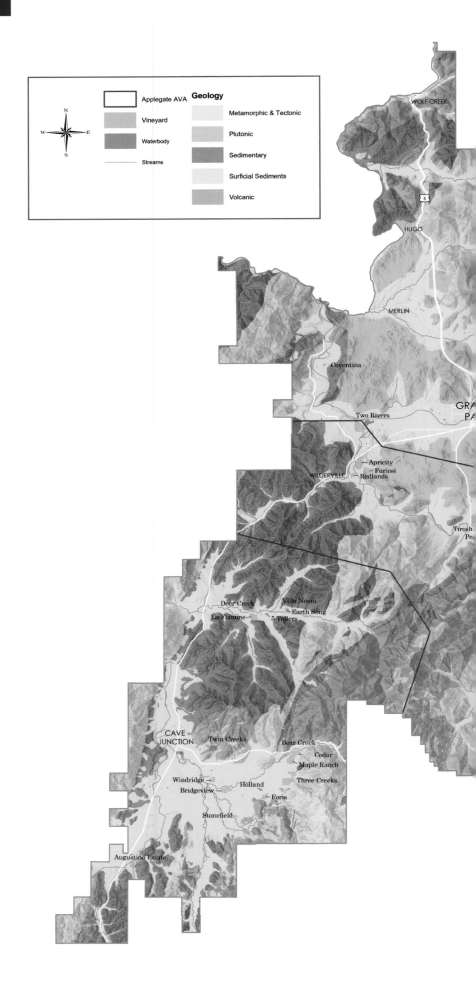

TRAIL

SHADY COVE

Somers Hardy

Folin

Cliff Creek

Kriselle

Buxton Ranch
Bayer Family Estates

EAGLE
POINT

Agate Ridge

Ousterhout

Venture

82

ROGUE RIVER

Del Rio

234

140

6

CENTRAL
POINT

Old Stage

Hummingbird Estate

Collar Family

Rocky Knoll

RoxyAnn

loro

ge & Sons

Serra

dt Family

Silver Stars

O'Nell

Layne

Rosella's

Conner

Wooldridge Creek

Rellik

Daisy Creek

Crater View

JACKSONVILLE

Dancin

MEDFORD

Grestoni

Dos Mariposas

Fences

Panner Hanson

Coal Mine

Peter William

Carlow

2 Hawk

Schultz

Walport Family

Troon

Sundown

Lakeside

Griffin
Creek

Eden
Vale

Carpenter Hill

PHOENIX

Jaxon

Stargazer
Estates

Rampart

Buren

APPLEGATE

Naumes

Pratt

La Luna

Luethy

Edgevale

Aardvark

Coleman Creek

Pebblestone

Chamberland

Padre Properties

Academy of Wines

Crooked Barn

Steelhead Run

Smitten Estate

Devitt

Jacksonville

Fly High

RUCH

Eevee's

Stoneriver

Aurora

Paschal

TALENT

Suncrest

Long Walk

Plaisance Ranch

Red Lily

Mae's
Longsword

Castellano

Stepping Stone

Gold Vineyard

Cedars

Pleasant Hill

Pleasant View

Dana Campbell

WILLIAMS

Becker

Valley View

Wells Land

Varner-Trail

Fortmiller

Tower View

Upper Five

ASHLAND

Quail Run

Swallow Hill

Gherky Creek

Grizzly
Peak

Bella Vista

plegate ValleyAVA

Cricket Hill

Eliana

Ashland

Pompadour

Belle Fiore

White Oak Hill

Irvine & Roberts

Cowhorn

Weisinger
Family

Rattlesnake Butte

John Michael
Champagne

Silver Stars

Phoenix

66

99

ENVIRONMENTAL RISK ANALYSIS FOR OIL AND GAS INDUSTRIES–LEASED PARCELS ACQUISITION PROCESS

Tetra Tech Inc.
Tampa, Florida, USA
By Gustavo Orozco

CONTACT
Gustavo Orozco
Gustavo.Orozco@tetratech.com

SOFTWARE
ArcMap 10.3, ArcScene 10.3, Model Builder 10.3, Microsoft Excel 2010

DATA SOURCES
Kansas Geological Survey, University of Kansas, ESRI, USGS NHD Dataset, State of Kansas GIS DASC, TES Blendspace, University of Missouri-Kansas City (UMKC)

This map shows a spatial analysis of human and environmental receptors that may be at risk due to oil and gas disasters. Risk analysis is an essential process to minimize the costs for environmental damage and social disturbances during the process of construction, extraction, and abandonment of oil and gas wells. The stage of leasing parcels for new oil and gas wells is one of the first steps toward the minimization of risks.

Countless disasters have damaged the environment and the quality of life for people. As a result, responsible oil companies have lost millions of dollars because of remediation, discredit, and depreciation of the stocks in the stock market.

For this research, an oil company needed to invest money in the acquisition of 20 percent of the available 268 leased parcels for well construction in Barber County: approximately 54 parcels. Because the financial investment was high and a single statistical analysis or a digital or paper map of the risks did not accurately represent all the complex relationships between a geographic space and its elements, the oil company needed a complete spatial analysis of the social and environmental receptors or risks to help it make an intelligent and accurate decision regarding what parcels it should lease. The project explores the analytic capacities of GIS and its multiple techniques, and it was developed in association with oil company experts and GIS analysts. The result was a list of each ranked parcel with minimal environmental and human risks to be provided to the permit personnel in the field.

Courtesy of TES Blendspace.

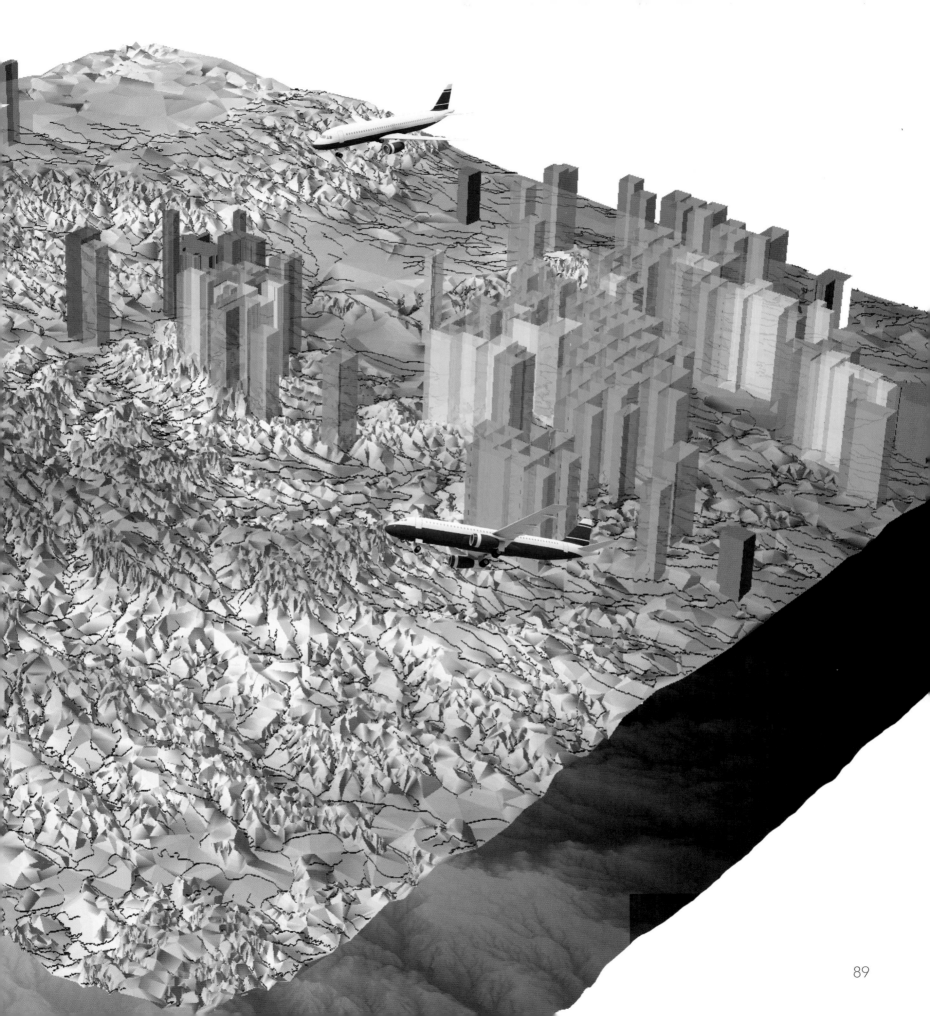

GEOLOGY OF THE NEWFIELD QUADRANGLE—CUMBERLAND, GLOUCESTER, AND SALEM COUNTIES, NEW JERSEY

NJ Department of Environmental Protection
Trenton, New Jersey, USA
By Michael V. Castelli and Zachary C. Schagrin

CONTACT
Helen Rancan
Helen.Rancan@dep.nj.gov

SOFTWARE
ArcGIS Desktop 10.6

DATA SOURCES
Buell, M.F. (1970); deVerteuil, Laurent (1997); Florer, L.E. (1972); Greller, A.M., Rachele L.D. (1983); Miller, K.G., Sugarman, P.J., et al (2001); Owens, J.P., Bybel, L.M., et al (1988); Rachele, L.D. (1976); Salibury, R.D., Knapp, G.N. (1917); Stanford, S.D. (2000); Stanford, S.D. (2015); Sugarman, P.J., Miller, K.G., et al (1993); Sugarman, P.J. and Monteverde, D. (2008)

This map shows the geology of the Newfield quadrangle mapped at a 1:24,000 scale, located in the Outer Coastal Plain of southern New Jersey. The surficial geology is shown as an overlay on a black and white USGS 1:24,000-scale quadrangle for Newfield, New Jersey, 1994 basemap. The underlying bedrock geology is shown in cross section. The map was created in ArcGIS by interpreting various layers such as historical, present day, and infrared aerial photography, lidar imagery, and digital elevation models (DEMs). All interpretations were ground-truthed by conducting fieldwork. There are also descriptions of the map units, a surficial deposit and geomorphic history of the map area, a correlation of the map units and figures that show geologic formations, and lidar interpretations in the map area. The attached table shows geologic interpretations for 251 domestic and public water supply wells in the Newfield quadrangle. This map is used for such purposes as water resource management, public health, engineering, mineral resources, and ecologic and historical resources by both public and private sectors. The map is available for download at the New Jersey Geological and Water Survey's website.

Courtesy of NJ Department of Environmental Protection.

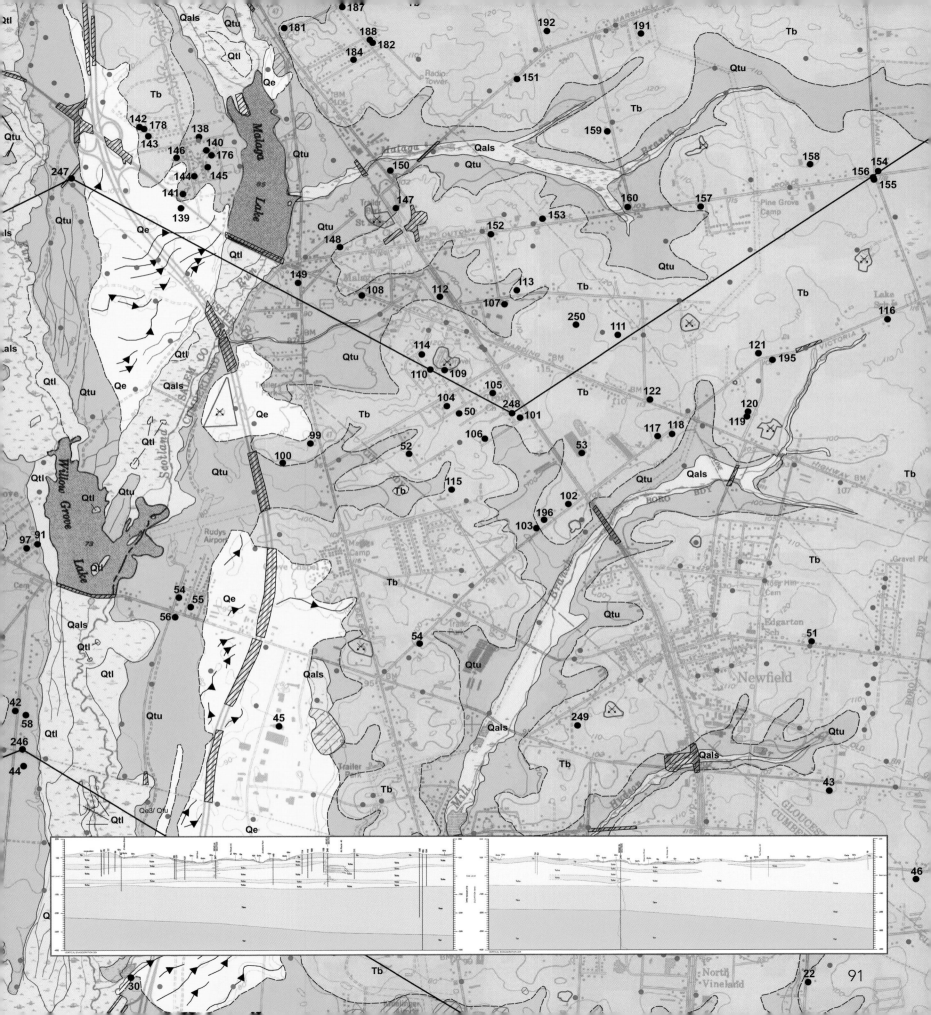

91

GEOLOGIC MAP OF THE HOMESTAKE RESERVOIR 7.5' QUADRANGLE– EAGLE, LAKE, AND PITKIN COUNTIES, COLORADO

U.S. Geological Survey
Denver, Colorado, USA
By Chester A. Ruleman, Michael G. Frothingham, Theodore R. Brandt, Colin A. Shaw, Marc W. Caffee, Keith A. Brugger, and Brent M. Goehring

CONTACT
Theodore Brandt
tbrandt@usgs.gov

SOFTWARE
ArcGIS Desktop 10.7.1, XTools Pro v19.0, Avenza MAPublisher v10.6.1

DATA SOURCES
US Geological Survey

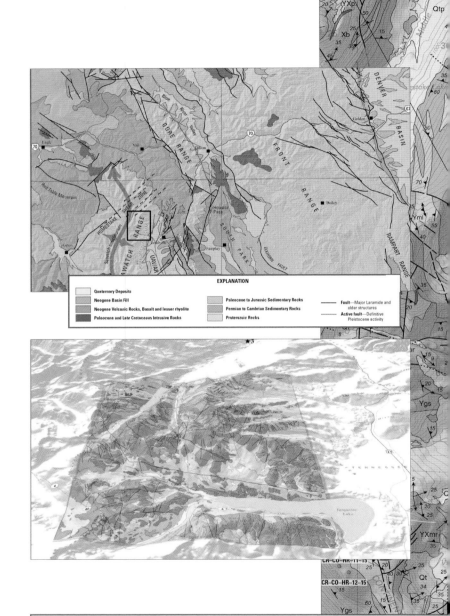

This map presents the geology of a 150-square-kilometer (58-square-mile) region straddling the Continental Divide at the northwestern end of the Upper Arkansas Valley, central Colorado. The region is popular with outdoor enthusiasts with access to several of Colorado's "Fourteeners," abundant wildlife, spectacular scenery, and fascinating geology. Elevations within the quadrangle range from 9,900 to 12,400 feet along the Continental Divide, forming the headwaters of the Arkansas and Colorado Rivers. The quadrangle lies within the Rio Grande rift, where active, intermountain-extensional faulting has down-dropped the Upper Arkansas Valley against the uplifted Sawatch Range. Ongoing uplift and middle Pleistocene glacial erosion have exposed a spectacular suite of ~1.8 to 1.4 billion-year-old metamorphic rocks that have subsequently been intruded by igneous rocks ranging in age from ~70 to 36 million years. These younger igneous rocks have created the heavily mineralized zones within this quadrangle and the Leadville region. Geochronologic data for the youngest glacial deposits reveal the position and longevity of the last glacial period, ~22,000 to 14,000 years ago. This is when glaciers reached their maximum, filling valleys as much as 300 meters (~1,000 feet) thick and subsequently melting back to snowfields along the Continental Divide.

In addition to data related to paleoclimate, seismic hazard, and mineral resources, we include accurately located mine shaft and tunnel data that can be used for hydrologic investigations related to basin structure and groundwater resources.

Courtesy of US Geological Survey.

THE HUMAN PRESENCE IN THE SOLAR SYSTEM

Space Research and Technology Institute,
Bulgarian Academy of Sciences
Sofia, Bulgaria
By Vanya Stamenova and Stefan Stamenov

It has been 50 years since the first time a human foot stepped on another celestial body. Since that time, humankind has sought to travel to other planets, satellites, asteroids, and even comets. It seems necessary and appropriate to incorporate our explorational endeavors into education, as we do with Earth's geography. Our human presence makes questions of conservation important, too. These maps show the locations of anthropogenic features, including spacecraft that have landed, intentionally impacted, or crashed on the surface of other planets, satellites, asteroids, and comets.

The map of the moon presents a global mosaic generated by wide-angle camera images from the Lunar Reconnaissance Orbiter (LRO) Camera, as well as anthropogenic features on the moon. The anthropogenic objects include soft-landed spacecraft, rovers, deployed scientific payloads, and spacecraft impact craters.

The map of Venus represents a colorized C3-MIDR mosaic developed to simulate the surface of Venus. The mosaic was created from the Magellan Full Resolution Basic Image Data Records (F-BIDRs), the highest resolution radar images of the Venus surface. More than 21 spacecraft have landed, intentionally impacted, or crashed on the Venus surface.

The map of Mars shows a digital elevation model (DEM) of Mars and the locations of anthropogenic features on its surface. The Mars anthropogenic features consist of spacecraft launched from Earth to explore it, including two spacecraft that are still operational—the Curiosity rover and InSight lander. The map also shows the estimated locations of about 14 missions on the surface of Mars.

The map of Mercury shows a colorized shaded relief of the original DEM of Mercury, generated using images acquired by the MESSENGER (Mercury Surface, Space Environment, Geochemistry, and Ranging) spacecraft, the first spacecraft to orbit Mercury.

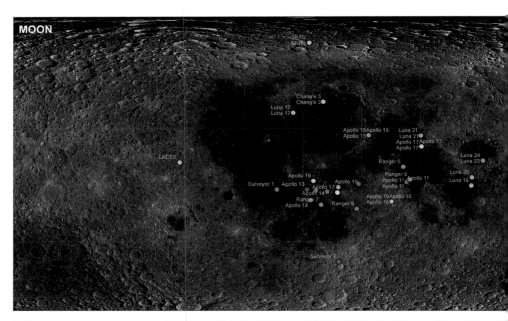

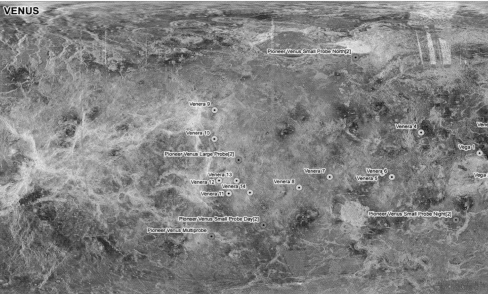

Courtesy of Dr. Vanya Stamenova and Dr. Stefan Stamenov.

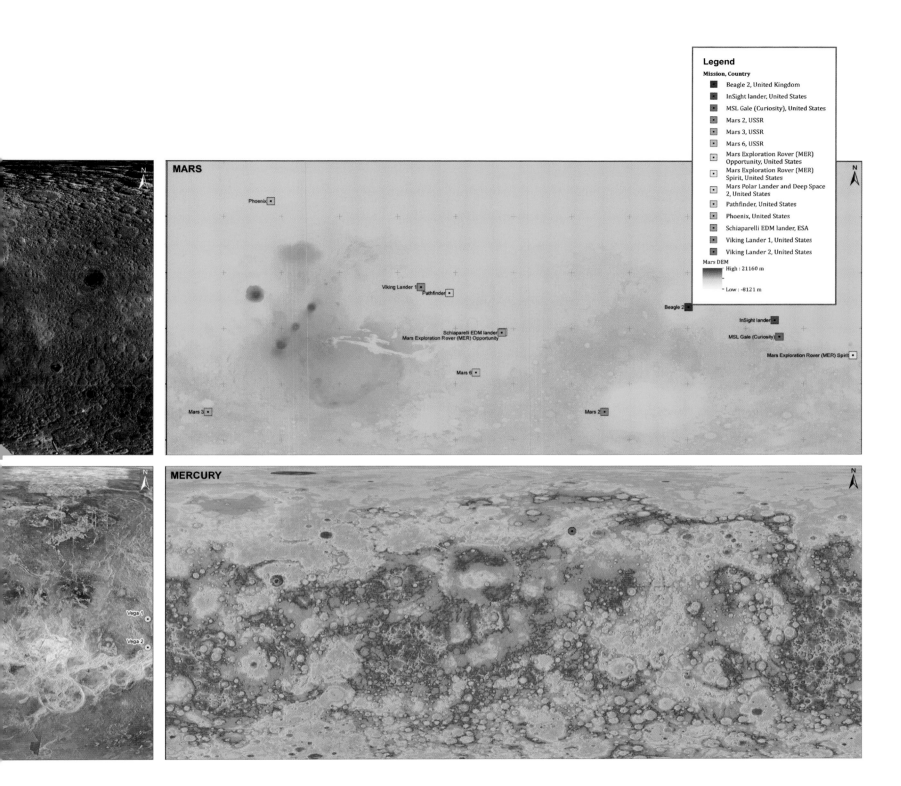

MARS

Phoenix

Viking Lander 1
Pathfinder

Schiaparelli EDM lander
Mars Exploration Rover (MER) Opportunity

Mars 6

Mars 3

Mars 2

Beagle 2

InSight lander

MSL Gale (Curiosity)

Mars Exploration Rover (MER) Spirit

Legend

Mission, Country

- Beagle 2, United Kingdom
- InSight lander, United States
- MSL Gale (Curiosity), United States
- Mars 2, USSR
- Mars 3, USSR
- Mars 6, USSR
- Mars Exploration Rover (MER) Opportunity, United States
- Mars Exploration Rover (MER) Spirit, United States
- Mars Polar Lander and Deep Space 2, United States
- Pathfinder, United States
- Phoenix, United States
- Schiaparelli EDM lander, ESA
- Viking Lander 1, United States
- Viking Lander 2, United States

Mars DEM

High : 21160 m

Low : -8121 m

MERCURY

Vega 1

Vega 2

CONTACT

Vanya Stamenova
vanya_stamenova@yahoo.com

SOFTWARE

ArcMap, ArcGIS Desktop 10.7, ECP grant to SCGIS Chapter Bulgaria

DATA SOURCES

NASA, GSFC, LROC website of Arizona State University, USGS Astrogeology Science Center (Planetary WMS service hosted by Astrogeology USGS and Lunar and Planetary Cartographic Catalog Astropedia at https://astrogeology.usgs.gov)

IDENTIFYING ACCURATE TOPOGRAPHIC DEPRESSIONS USING QUALITY LEVEL 2 LIDAR DATA

Ayres Associates Inc.
Madison, Wisconsin, USA
By Bruce Riesterer and Adam Derringer

CONTACT
Adam Derringer
DerringerA@AyresAssociates.com

SOFTWARE
ArcGIS Pro 2.5, ArcGIS Spatial Analyst™

DATA SOURCES
Door County Wisconsin, Ayres Geospatial, Esri

This analytic map displays stages of topographic depression extraction using high-quality lidar data and a comprehensive countywide road culvert location inventory.

Road culvert locations facilitate the proper routing of water flows across the landscape in hydrologic modeling by correcting the digital dam effect of roadbeds in lidar surface elevation data. When water flows are modeled properly, true topographic depressions can be extracted. After topographic depressions are isolated, detailed analyses of individual sites can be performed, including depression low-point identification, depression maximum depths, and depression void-volume calculations. Planning and zoning professionals value accurate locations and depths of depressions for assessing flood potential, analyzing water quality and quantity concerns, implementing wetland protection measures, and evaluating land development limitations.

This map incorporates a lidar digital elevation model (DEM) raster image, a topographic depression raster image with depth shading, a topographic depression polygon layer, and a 3D view of topographic depressions on a small segment of Door County, Wisconsin.

Courtesy of Ayres Associates.

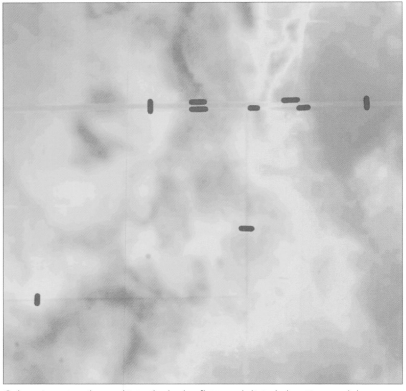

Culvert inventory burned into the hydro-flattened digital elevation model to produce a hydro-enforced digital elevation model.

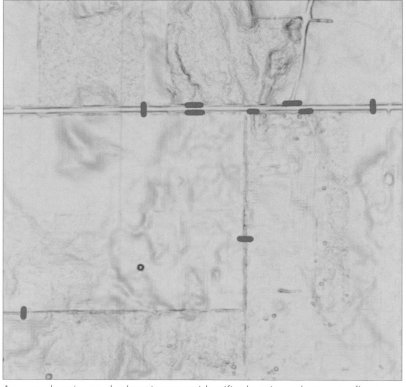

A comprehensive road culvert inventory identifies locations where water flow paths are altered by roadbed fill in hydrological modeling.

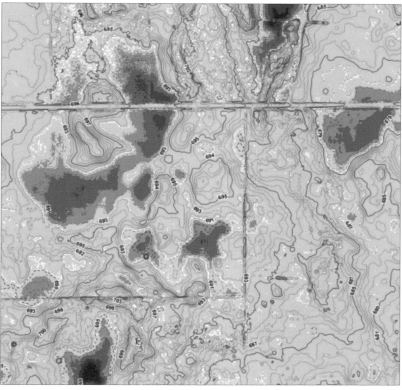

Extracted depressions using hydrology tools with symbolization to indicate depth ranges. Contours enhance the level of detail of the mapped depressions.

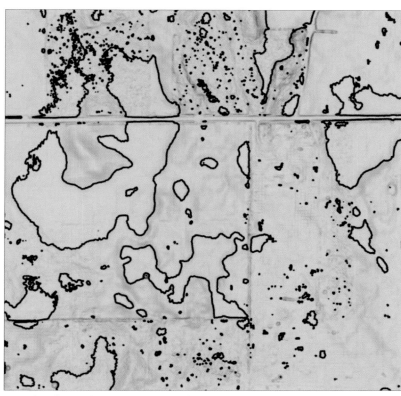

Boundary feature layer generated at the perimeter of each depression. Multiple layer formats provide for increased levels of data analysis.

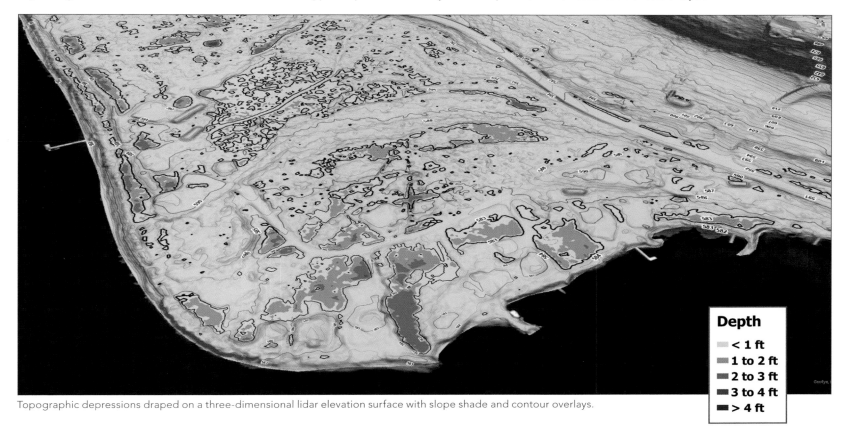

Topographic depressions draped on a three-dimensional lidar elevation surface with slope shade and contour overlays.

Depth

- < 1 ft
- 1 to 2 ft
- 2 to 3 ft
- 3 to 4 ft
- > 4 ft

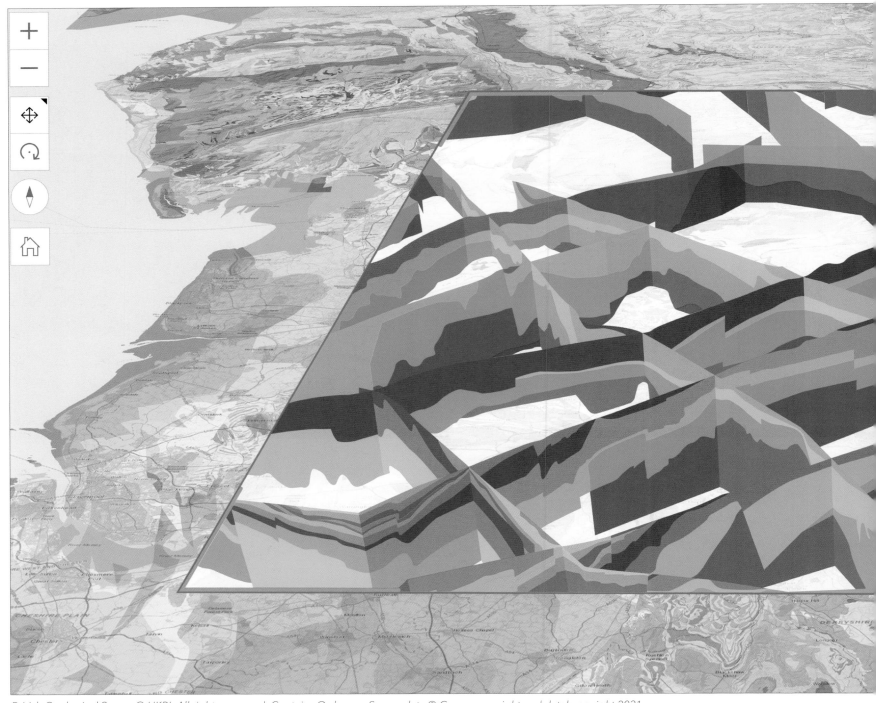

VIEWING THE UK'S SUBSURFACE WITH THE ARCGIS API FOR JAVASCRIPT

British Geological Survey
Keyworth, Nottingham,
United Kingdom
By Andy Marchant and
Henry Holbrook

This map shows the British Geological Survey (BGS) UK3D dataset. UK3D is a national-scale network of intersecting cross sections (also known as a fence diagram), providing a unique visualization of the complex rocks and structures that make up the UK landmass. The cross sections have been used by BGS and partner organizations to develop national-scale understanding of aquifers and groundwater flow and to inform screening processes for identifying potential sources of energy and sites for waste storage.

The map was made using the BGS 3D Geology of Britain application. Built using the ArcGIS API for JavaScript,

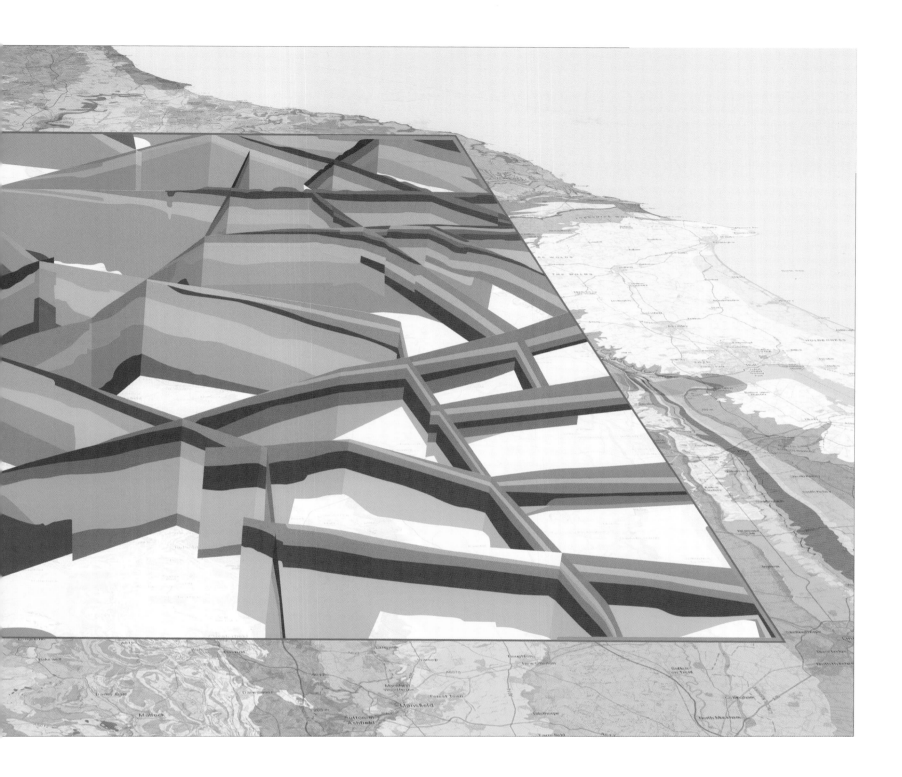

the application allows users to view the UK's surface geology draped over an elevation surface. The Esri Slice widget was used to cut through the surface to view the cross sections beneath.

CONTACT
Andy Marchant
a.marchant@bgs.ac.uk

SOFTWARE
ArcGIS Online, ArcGIS API v4.16 for JavaScript

DATA SOURCES
British Geological Survey: 1:50,000 scale geology data and UK3D cross sections

ASSESSMENT OF SUSCEPTIBILITY TO LANDSLIDES THROUGH NEURAL NETWORKS NEAR SIMON BOLIVAR AVENUE, ECUADOR

Universidad de las Fuerzas Armadas ESPE
Sangolquí, Pichincha, Ecuador
By Andrés Betancourt, Andrea Castro, and Oswaldo Padilla

CONTACT
Andrés Betancourt
andresbetancourt1995@gmail.com

SOFTWARE
ArcGIS Desktop 10.7

DATA SOURCES
University of the Armed Forces ESPE, National System of Information and Management of Rural Lands and Technological Infrastructure SIGTIERRAS, Military Geographic Institute IGM, Public Metropolitan Company of Potable Water and
Sanitation of Quito EPMAPS

Landslides are one of the most frequent threats in the Metropolitan District of Quito (MDQ) in Ecuador. They generally occur in anthropic areas, such as the road network, which affects transportation, the economy, and above all, citizen security. Therefore, this study offers support to evaluate the susceptibility to landslides in the areas near Simon Bolivar Avenue, a very affected area. The analysis is carried out based on the neural network algorithm (NNA) methodology, which developed a model focused on the previous landslides registered, connecting them with influential factors on the phenomenon as independent variables (slope, precipitation, vegetal coverage, lithology, distance to geologic faults, rivers, and road network). They were standardized, weighted, and classified according to the landslide level (high, medium, and low).

The susceptibility mapping for this methodology came from the training and validation of a multilayer perceptron network, which was tested with different layers hidden, each with its corresponding number of neurons. The network that is most adaptable to the data is the one that has two hidden layers with 19 and 7 neurons, respectively, obtaining from it the high, medium, and low landslide susceptibility zones from the research area.

Courtesy of University of the Armed Forces ESPE

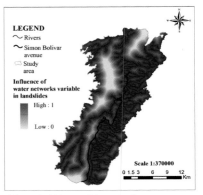

Map 1: Water Networks Variable Standardization

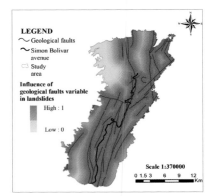

Map 2: Geological Faults Variable Standardization

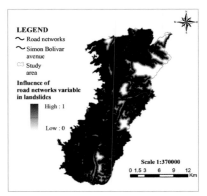

Map 3: Road Networks Variable Standardization

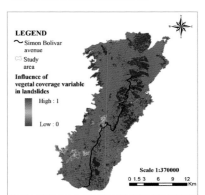

Map 4: Vegetal Coverage Variable Standardization

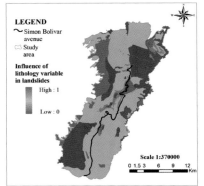

Map 5: Lithology Variable Standardization

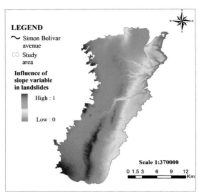

Map 6: Slope Variable Standardization

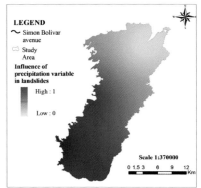

Map 7: Precipitation Variable Standardization

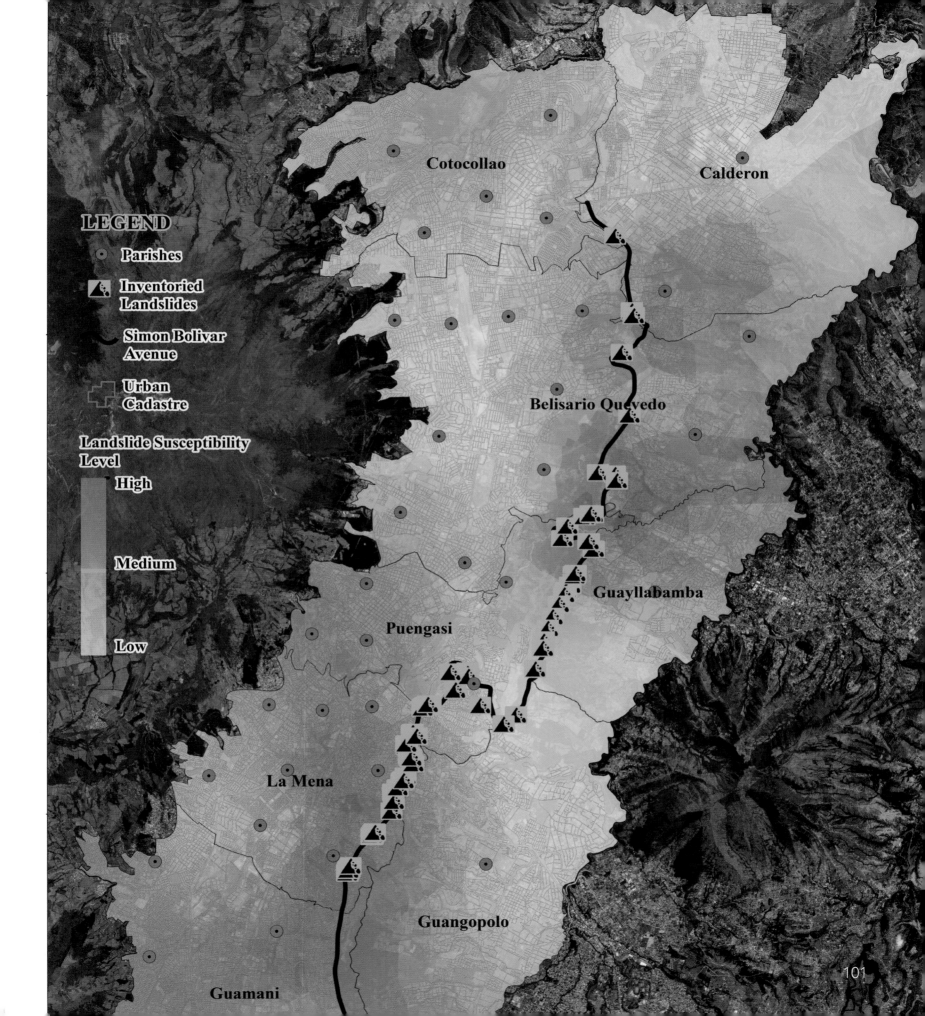

LEGEND

⊙ Parishes

◭ Inventoried Landslides

— Simon Bolivar Avenue

⌐ Urban Cadastre

Landslide Susceptibility Level

High

Medium

Low

Cotocollao

Calderon

Belisario Quevedo

Guayllabamba

Puengasi

La Mena

Guangopolo

Guamani

101

VISUALIZING SEISMIC AND WELL DATA IN ARCGIS PRO

Exprodat Consulting Ltd.
London, England, United Kingdom
By Richard Webb, Chris Jepps, Ross Smail, and David Wilton

CONTACT
Exprodat
info@exprodat.com

SOFTWARE
ArcGIS Pro

DATA SOURCES
Exprodat

Seismic surveys use reflected sound waves to produce a scan of the earth's subsurface. The seismic data produces a key tool for oil and gas exploration by helping geoscientists identify and delineate structures below the surface that could contain petroleum. From seismic surveys, oil and gas operators can identify suitable locations for drilling wells to probe the subsurface and test for hydrocarbons. When drilling, operators use wireline logging to obtain a continuous record of a formation's rock properties, by running instruments down the hole that has been drilled. From the geophysical logs produced, operators can deduce key information about the subsurface geology, such as lithology, porosity, and permeability.

The image provided shows an ArcGIS Pro 3D scene containing the wireline logs from a well drilled in the offshore Netherlands sector of the North Sea, against a backdrop of 3D seismic survey data. The seismic data was imported from SEG-Y format and the well log data from industry-standard LAS format, both using Data Assistant and imported directly into ArcGIS Pro.

Courtesy of Exprodat Consulting Ltd.

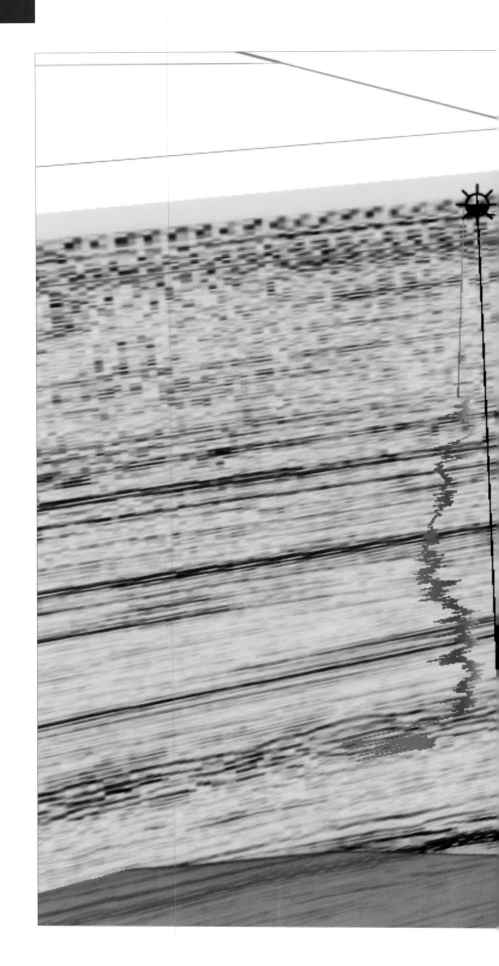

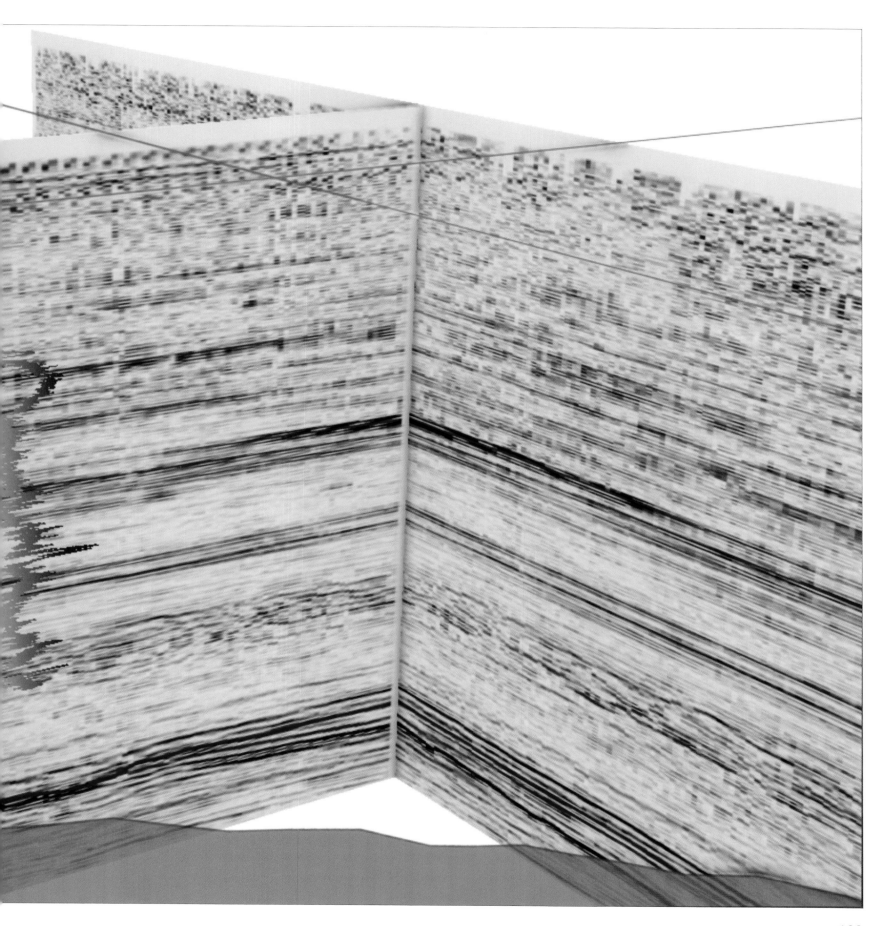

UNITED STATES FOREST SERVICE SUBSECTION ECOLOGICAL UNITS

USDA-USFS-GTAC
Salt Lake City, Utah, USA
By Claire Simpson, Larry Laing, Robert (Andy) Colter, and Greg Nowacki

CONTACT
Nathan Pugh
nathan.pugh@usda.gov

SOFTWARE
ArcGIS Pro, ArcGIS Online

DATA SOURCES
DOI, USDA-USFS

The US Department of Agriculture maintains Terrestrial Ecological Unit Inventory (TEUI) data consisting of multiple landscape and land unit scaled ecological types. This data is used in natural resource inventory, monitoring, and land management planning, as well as to make predictions and interpretations of National Forest System lands. Subsection units fall in the subregion scale of TEUI data; these areas are characterized by similar regional climate, geomorphic process, topography, and stratigraphy. This map illustrates the latest USFS updates to subsection polygons for the contiguous United States overlain on soil texture (surface-level percent sand, silt, and clay are displayed as the RGB bands to show relative proportions), one of the data layers that influence subsection boundaries. To create subsections, experts, guided by knowledge of region-specific ecological drivers, assess a suite of geospatial data layers and manually delineate subsections to capture dominant landscape patterns. Subsections range in size from <1 acre to >10 million acres, where size is inversely proportional to climatic, topographic, and geologic landscape variability. For example, this map shows large subsections in the flatter, more homogeneous Midwest region and small subsections in the dynamic Western United States. Ultimately, subsections are characterized by zonal statistics, which allow for quantitative upward aggregation to create a nested national hierarchy of land units.

Courtesy of A. Ramcharan, T. Hengl, T. Nauman, C. Brungard, S. Waltman, S. Wills, and J. Thompson 2018. Soil property and class maps of the conterminous United States at 100-meter spatial resolution. Soil Science Society of America Journal.

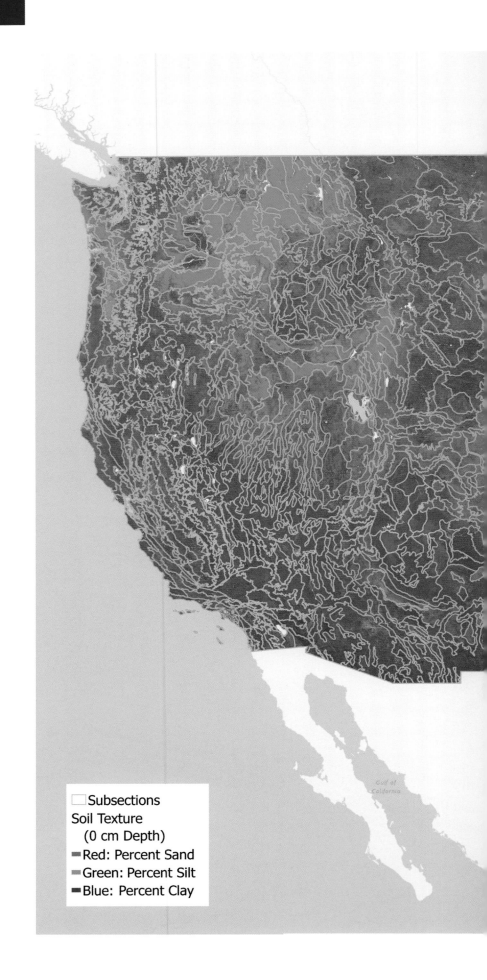

Subsections
Soil Texture
(0 cm Depth)
Red: Percent Sand
Green: Percent Silt
Blue: Percent Clay

St Lawrence Seaway

Bay of Fundy

THE BAHAMAS

CUBA

Gulf of Mexico

Gulf of Mexico

Straits of Florida

MÉXICO

Yucatan Channel

Esri, HERE, NPS

EARTH'S CRUST–EXAGGERATED TOPOBATHY ELEVATIONS

Teck Resources
Vancouver, British Columbia, Canada
By Michael Kelly

CONTACT
Michael Kelly
Michael.Kelly@teck.com

SOFTWARE
ArcGIS Online, ArcGIS Living Atlas of the World,
ArcGIS API 4.15 for JavaScript

DATA SOURCES
ArcGIS Living Atlas of the World

A 3D web mapping application illustrating exaggerated topography and bathymetry (TopoBathy).

The earth is far from flat, and this includes the ocean floor, which is hidden underwater. This application opens the ocean floor to everyone and allows users to explore hidden mountains by exaggerating the elevations of all surfaces. Add ArcGIS Living Atlas of the World layers of interest to this app, and you can see the world in a very different light.

Courtesy of Teck Resources.

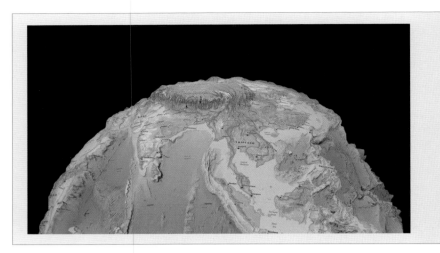

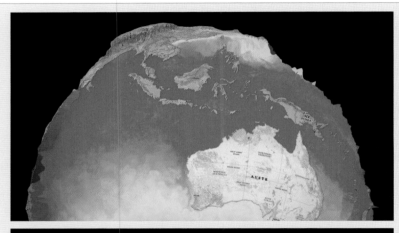

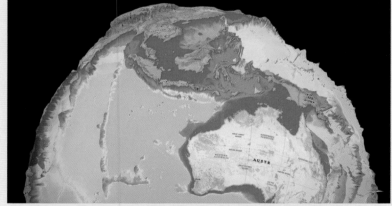

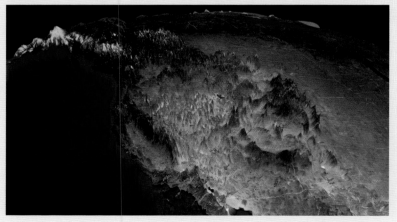

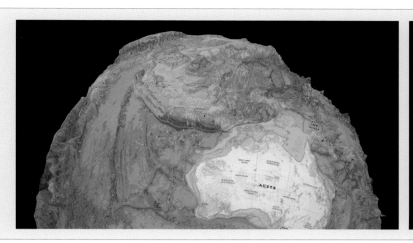
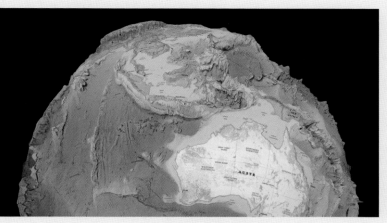
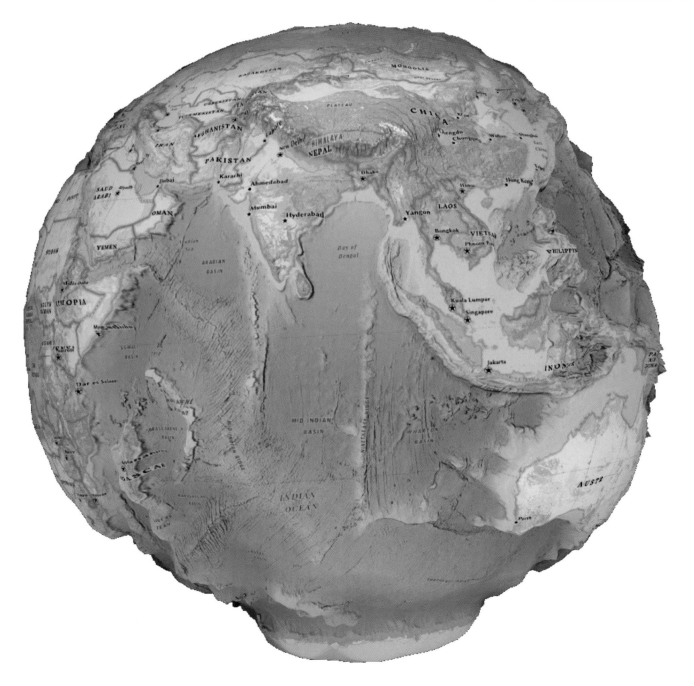

LARGE SUBMARINE CANYONS OF THE US OUTER CONTINENTAL SHELF

CSA Ocean Sciences Inc., Stuart, Florida, USA

By Melanie L. Cahill, Stephen T. Viada, Mark S. Fonseca, Steve W. Ross, Fabio C. De Leo, Dustin Myers, Brian Diunizio, Brent Gore, Charles Hagens, Emille Rodriguez, Kristen L. Metzger, Ashley A. Lawson, Keith Vangraafeiland, Ankur Patel, and Paul O. Knorr

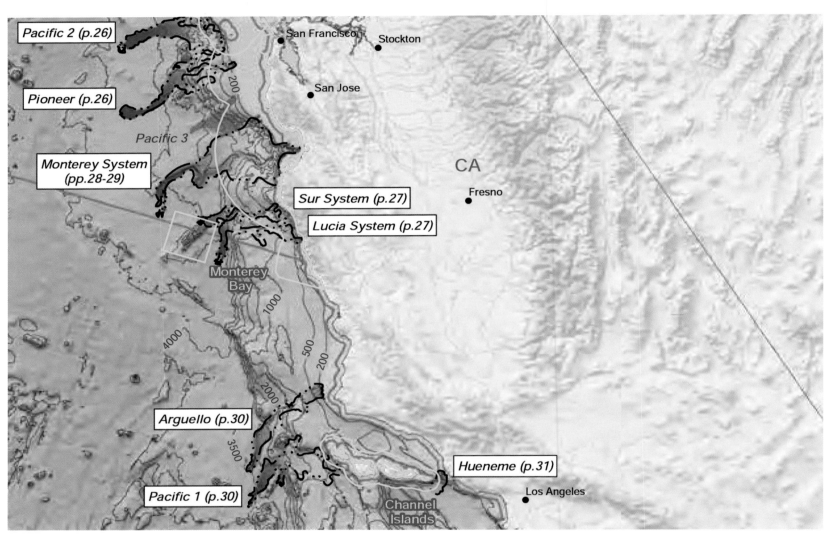

Courtesy of CSA Ocean Sciences, Inc.

The purpose of this atlas is to facilitate improved environmental management of the Outer Continental Shelf (OCS) by developing one depository of maps and information on major submarine canyons of the OCS. This atlas provides the Bureau of Ocean Energy Management (BOEM) with geospatial and resource information to assist in the preparation of environmental documents. To accomplish this, submarine canyons were inventoried and delineated using a methodology consistent with terrestrial watershed mapping. A criteria-based algorithm generated spatial polygons used to calculate canyon slope, length, and depth. A concurrent literature review was conducted, which provided the notable facts seen in the atlas. In addition, the literature citations were cataloged in an EndNote library and a synopsis was embedded in the geodatabase files.

Canyon polygons developed by Harris et al. (2014) were the starting point for defining canyon boundaries and guiding the initial canyon inventory and selection process. Harris used both GIS tools (Topographic Position Index, calculated by comparing each cell slope value to the mean slope of the cell's neighbors) and the judgment of subject matter experts.

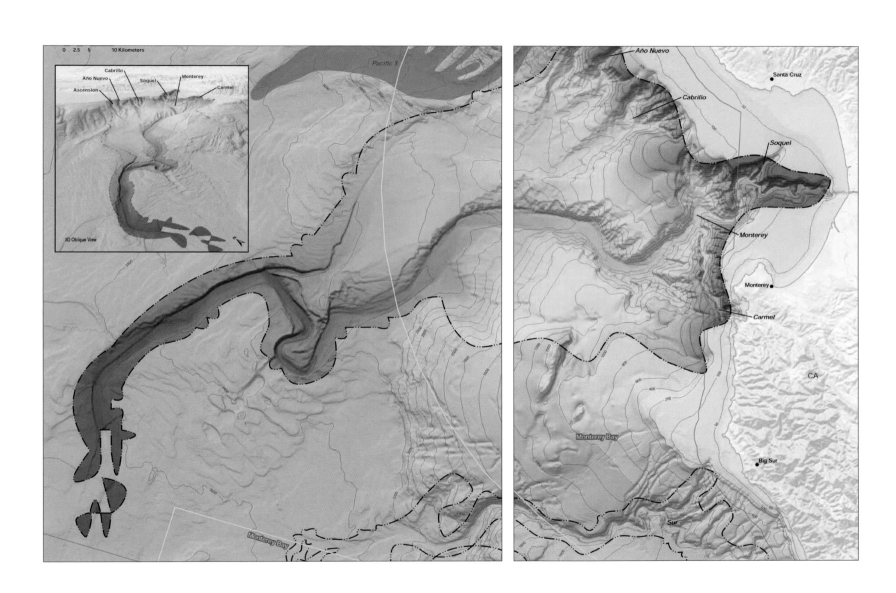

For this project, CSA Ocean Sciences Inc. developed an objective, numerically repeatable delineation process using the Harris data as a starting point. As with Harris, canyon bathymetry and boundaries were derived using Shuttle Radar Topography Mission (SRTM30) data because of its consistent geographic coverage.

CONTACT
Dustin Myers
dmyers@conshelf.com

SOFTWARE
ArcGIS Desktop 10.6

DATA SOURCES
Harris et al., CSA Ocean Sciences Inc.

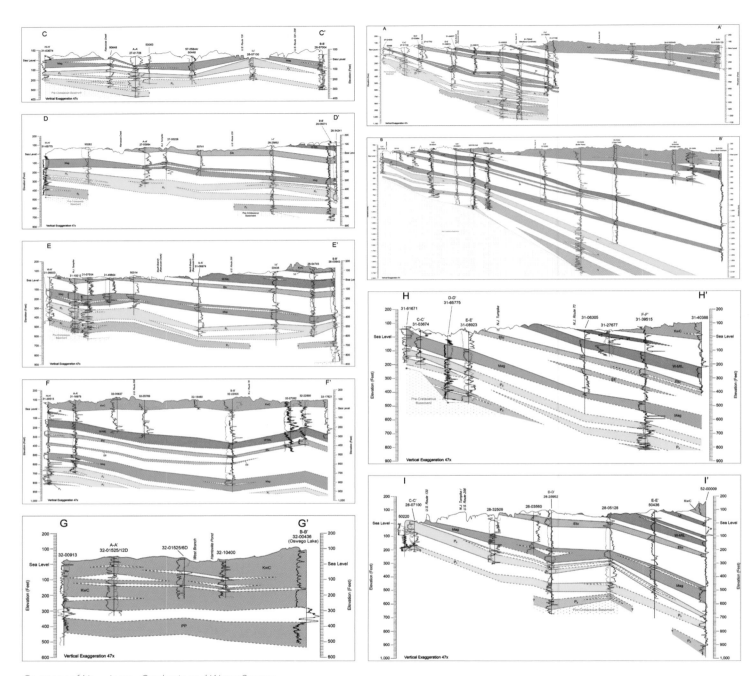

Courtesy of New Jersey Geologic and Water Survey.

FRAMEWORK AND PROPERTIES OF AQUIFERS IN BURLINGTON COUNTY, NEW JERSEY

NJ Department of Environmental Protection
Trenton, New Jersey, USA
By Sugarman, Carone, Stroiteleva, Pristas, Monteverde, Domber, Filo, Rea, and Schagrin

This map illustrates the thickness and extent of aquifers and confining units in Burlington County, New Jersey. The map also provides the hydrogeologic properties of these aquifers.

CONTACT
Helen Rancan
Helen.Rancan@dep.nj.gov

SOFTWARE
ArcGIS Desktop 10.6

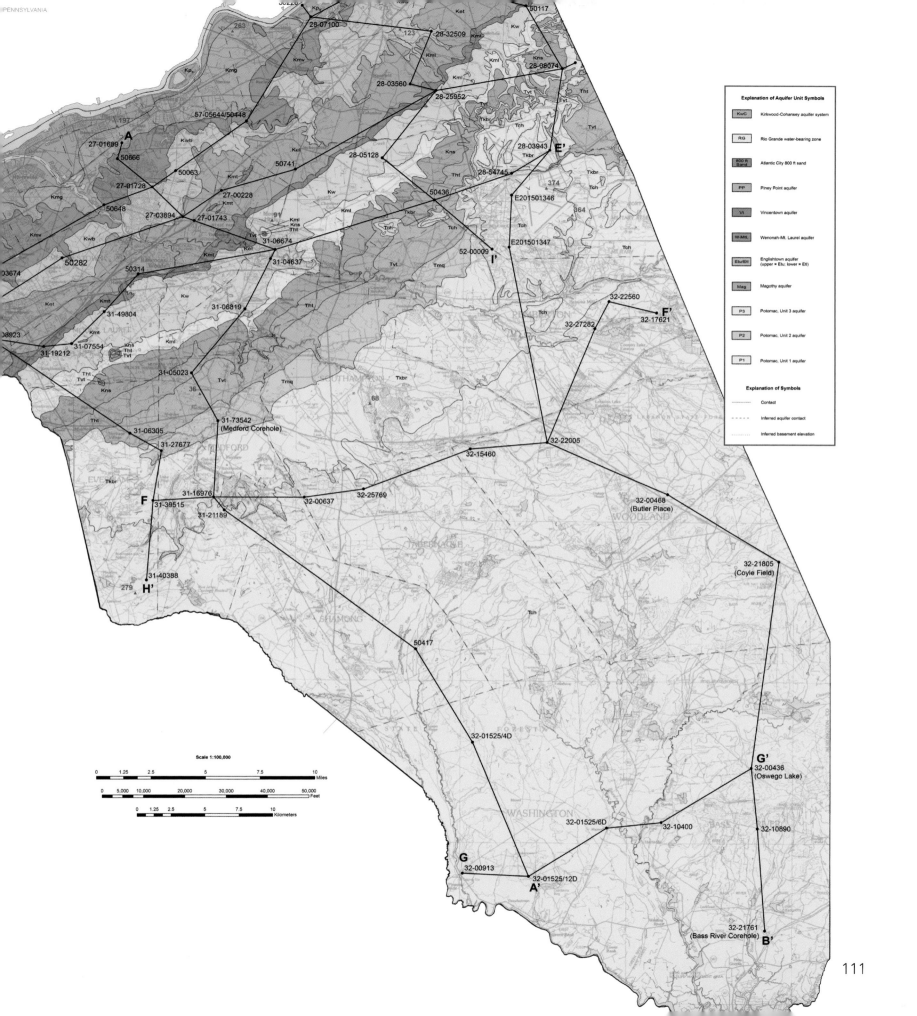

ICELANDIC COLD WATER SPRINGS

Iceland GeoSurvey
Reykjavik, Iceland
By Albert Thorbergsson and Árni Hjartarson

CONTACT
Albert Thorbergsson
albert.thorbergsson@isor.is

SOFTWARE
ArcGIS Desktop

DATA SOURCES
Iceland GeoSurvey, National Land Survey of Iceland

This map gives an overview of the freshwater springs in Iceland. Because of high annual precipitation and permeable bedrock formations, Iceland is rich in groundwater, and springs are common. Their distribution is, however, uneven. Springs are primarily found inside the active volcanic belts but are rarer in the older regions of the country. The world's largest springs and spring areas are found in Iceland, and they issue thousands of liters per second. They are mainly connected to young volcanic formations, Holocene lavas, and active fissure swarms. These springs are characterized by low chemical concentration, even runoff, and more or less constant water temperature all year around. Outside the volcanic belts, the springs are smaller and more fluctuating. On the map, springs and spring areas are categorized due to their discharge into five classes (see the legend). In Iceland, large areas of no perennial surface runoff exist. These areas are also indicated on the map.

Courtesy of ÍSOR-Iceland GeoSurvey.

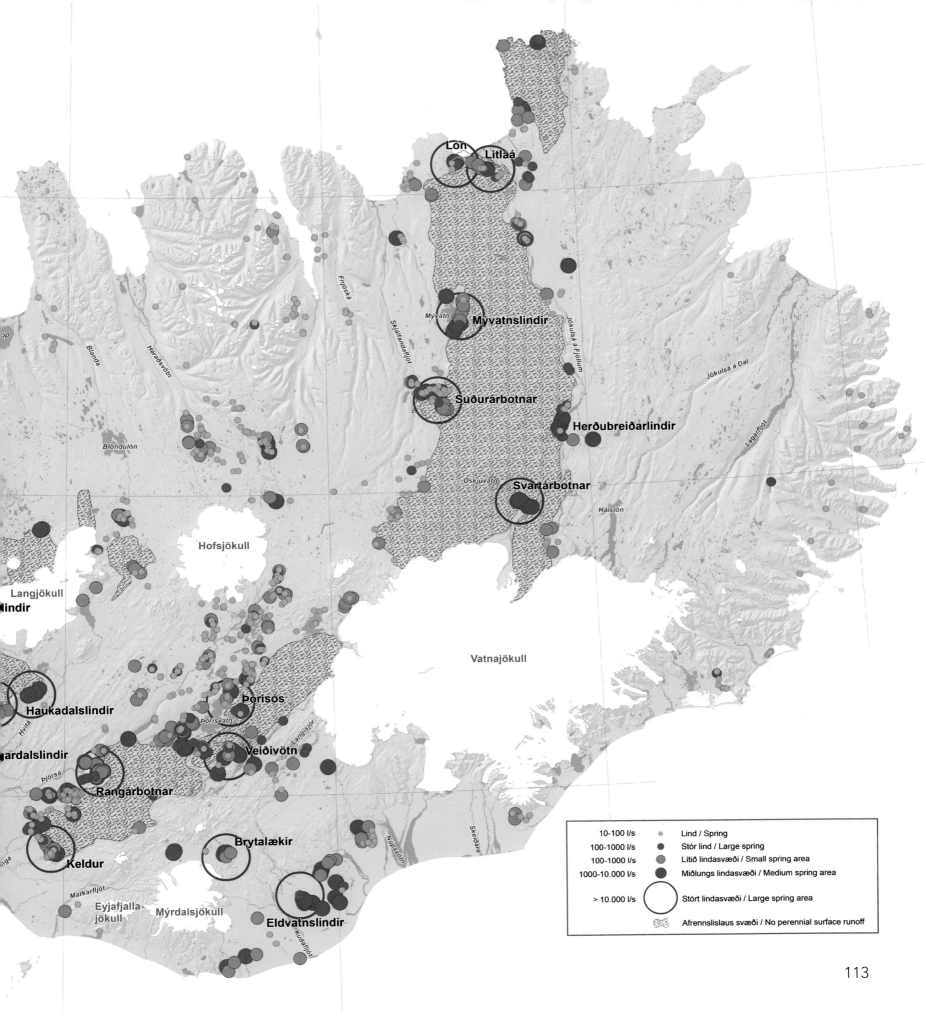

Lón
Litlaá
Mývatnslindir
Suðurárbotnar
Herðubreiðarlindir
Svartárbotnar

Hofsjökull

Langjökull
lindir

Vatnajökull

Þórisós

Haukadalslindir
ardalslindir
Veiðivötn

Rangárbotnar

Brytalækir

Keldur

Eyjafjalla-
jökull
Mýrdalsjökull

Eldvatnslindir

Blanda
Héraðsvötn
Fnjóská
Skjálfandafljót
Mývatn
Jökulsá á Fjöllum
Jökulsá á Dal
Lagarfljót
Öskjuvatn
Hálslón

Blöndulón

Hvítá
Þjórsá
Þórisvatn
Langisjór
Núpsvötn
Skeiðará

Markarfljót
Kúðafljót

10-100 l/s	○	Lind / Spring
100-1000 l/s	●	Stór lind / Large spring
100-1000 l/s	●	Lítið lindasvæði / Small spring area
1000-10.000 l/s	●	Miðlungs lindasvæði / Medium spring area
> 10.000 l/s	◯	Stórt lindasvæði / Large spring area
	🌀	Afrennslislaus svæði / No perennial surface runoff

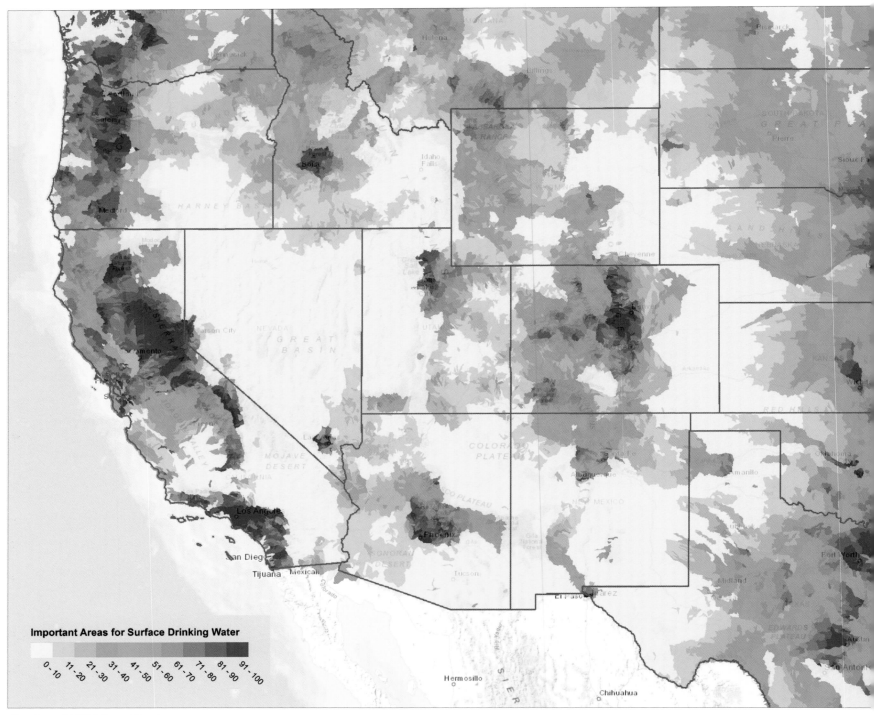

Important Areas for Surface Drinking Water

0 - 10 11 - 20 21 - 30 31 - 40 41 - 50 51 - 60 61 - 70 71 - 80 81 - 90 91 - 100

Courtesy of USDA Forest Service.

FORESTS TO FAUCETS 2.0

US Forest Service
Durham, North Carolina, USA
By Rebecca Lilja, Erika Mack, Cass Klee, and Michelle Hawks

Forests to Faucets 2.0 (F2F2) assesses subwatersheds (HUCS12) in the United States to identify those that are important to downstream surface drinking water supplies as well as evaluate a watershed's natural ability to produce clean water. F2F2 updates Forests to Faucets v1 (2011) and adds new future risks to watershed, such as climate-induced changes in land use and water quantity.

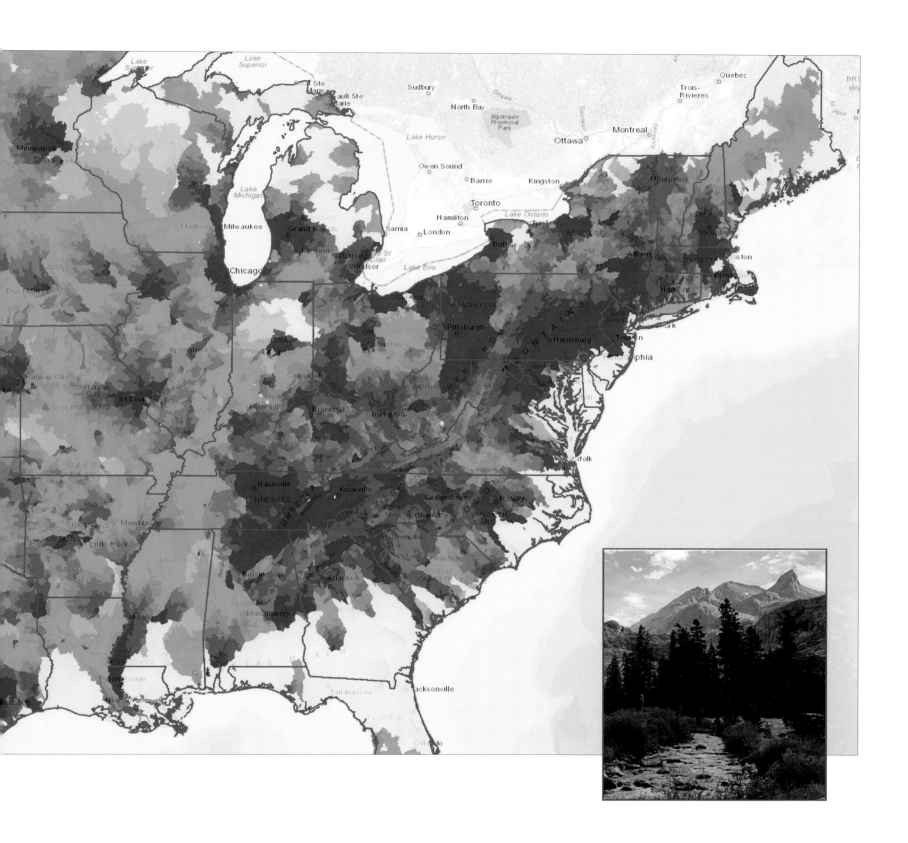

CONTACT
Rebecca Lilja
rebecca.l.lilja@usda.gov

SOFTWARE
ArcGIS Pro, ArcGIS Online

DATA SOURCES
Forests to Faucets 2.0

USFS–CHUGACH NATIONAL FOREST LAKES AND RIVERS

USDA–United States Forest Service
Salt Lake City, Utah, USA
By Nick Klein-Baer, Kim Homan, and Nathan Pugh

CONTACT
Nathan Pugh
nathan.pugh@usda.gov

SOFTWARE
ArcGIS Pro

DATA SOURCES
USFS resource imagery, IFSAR, Landsat, Sentinel, Esri World Imagery

The US Department of Agriculture, Forest Service is working to update and modernize hydrography data for Forest Service lands in the state of Alaska. The USGS National Hydrography Dataset (NHD) for Alaska was derived from 1950s 1:63,360 scale topographic maps. At the time of completion in 2008, much of it fell short of national high-resolution standards. Since then, the Alaska Hydrography Technical Working Group (AHTWG) has led a cooperative effort toward producing a unified statewide hydrography product, completing updates for more than 20 percent of the state's hydrography.

These maps show the results (highlighted in red) of a semiautomated process developed to delineate 2D water features according to NHD and AK Hydrography Database standards, as applied to two watersheds surrounding Prince William Sound on the Chugach National Forest in Alaska. Advanced image processing techniques—including object-based image analysis and machine learning algorithms—were applied to high-resolution satellite imagery and topographic data derived from airborne synthetic aperture radar (SAR) to detect lakes and ponds greater than or equal to two acres in size, as well as streams and rivers wider than 20 meters. The output data was then uploaded into the NHD and made available for use to other state and federal agencies and the public. Additional 2D water feature mapping is currently being completed on the Chugach and Tongass National Forests in Alaska.

Courtesy of USDA–United States Forest Service.

Legend
NHD 2D Lake and River Features

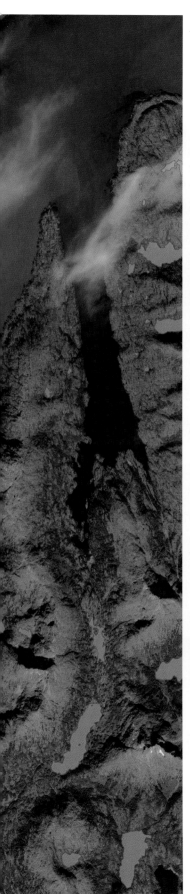

CHARLESTON PENINSULA STUDY: COASTAL STORM RISK MANAGEMENT STUDY FOR CHARLESTON, SC

USACE Charleston District
Charleston, South
Carolina, USA
By Jennifer Kist

The Charleston Peninsula Study is a three-year, $3 million supplemental study undertaken by the US Army Corps of Engineers (USACE) to assess the threat of coastal storm surge to historic downtown Charleston, South Carolina. A key report in the study was scheduled for public release right at the peak of COVID-19, so the USACE had to determine a way to engage the public on an extensive technical report without physical contact.

Cultural Resources

Without Project 2082
- 119 (54%) historic structures at risk
- 1 square mile (51%) of historic area at risk
- 0.1 square miles (43%) of archaeology sites at risk

With Project 2082
- 5 (2%) historic structures at risk
- 0.2 square miles (7%) of historic area at risk
- 0.02 square miles (13%) of archaeology sites at risk

Courtesy of USACE Charleston District.

The Peninsula Study Story Map was developed to give the public an interactive walk-through of the draft report for the Charleston Peninsula Study. The report has garnered over 20,000 views and resulted in almost 500 public comments on the draft report.

CONTACT
Jennifer Kist
Jennifer.K.Kist@usace.army.mil

SOFTWARE
ArcGIS Pro 2.4.2 and
ArcGIS Online

DATA SOURCES
USACE Charleston District

TRANSPORTATION SAFETY ANALYSIS: OTTAWA, ONTARIO

Cardinal Geospatial, Chapel Hill, North Carolina, USA
By Alex Reinwald and Brett Cox

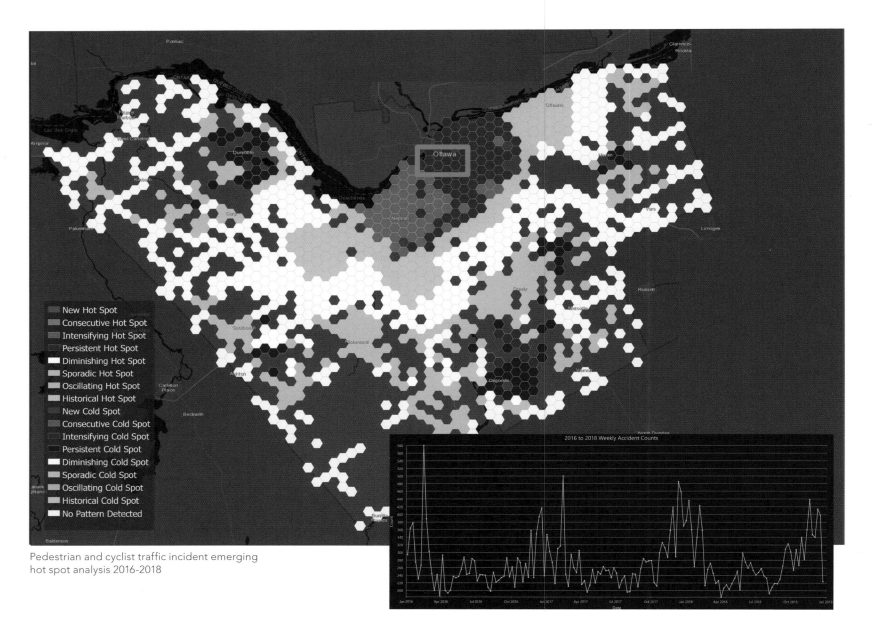

Pedestrian and cyclist traffic incident emerging
hot spot analysis 2016-2018

The charts included in this map illuminate how the time of year affects the number of traffic incidents that occur. Most apparent when looking at these charts is the heightened intensity during the winter. The eight most problematic dates occur from November to March, with a clear correlation of higher incident numbers overall during these months. These trends are proven over the course of three years in the line graph.

Taken further, data that includes time can begin to show temporal trends. The temporal hot spot analysis for the City of Ottawa not only outlines areas that are hot and cold spots when it comes to traffic incidents. It also provides visibility into how the number of incidents is trending year to year. Near Parliament Hill, we see that incidents are reliably high year over year. In the Nepean area, incident numbers have been reliably high but are intensifying.

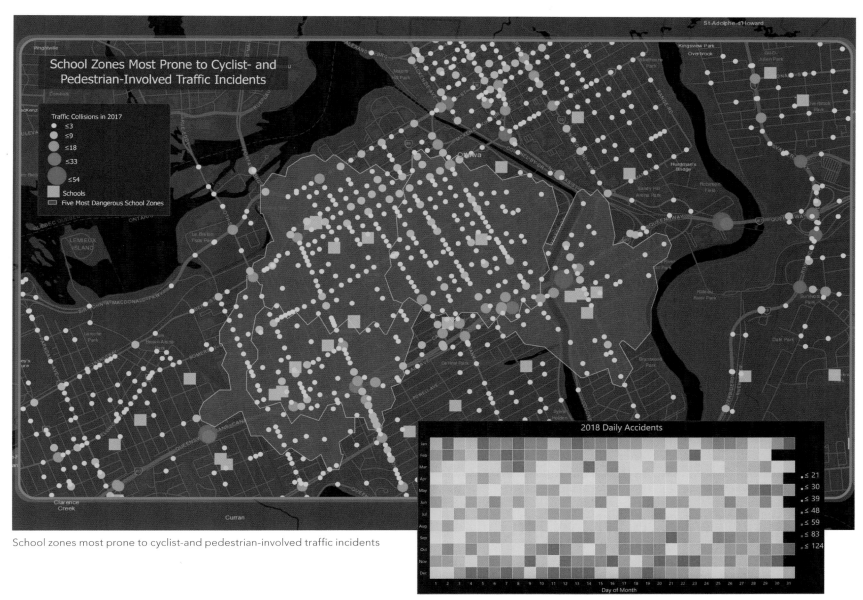

School zones most prone to cyclist-and pedestrian-involved traffic incidents

Courtesy of Alex Reinwald and Brett Cox of Cardinal Geospatial.

Another priority for public safety officials is increasing safety in school zones, especially in the high-intensity area of Parliament Hill. Identifying school zones with a disproportionate number of incidents will aid in targeting marketing efforts for school transportation safety. The inset displays the five school zones most prone to traffic incidents.

CONTACT
Alex Reinwald
alexereinwald@gmail.com

SOFTWARE
ArcGIS Pro 2.5

DATA SOURCES
City of Ottawa

Courtesy of Bayerisches Staatsministerium für Wohnen, Bau und Verkehr.

AI OBJECT RECOGNITION FROM LIDAR DATA

Bayerisches
Staatsministerium für
Wohnen,
Bau und Verkehr
Munich, Germany
By Roland Degelmann

Comprehensive knowledge of the existing transport infrastructure is an essential basis for planning, construction, and usage processes. In addition to traditional surveying methods, kinematic 3D laser scan measurement is increasingly used within modern asset management. This method allows a fast and highly accurate recording of the road space in the form of 3D point clouds. But the derivation of computer-assisted design (CAD) objects from these point clouds is still time-consuming and expensive. The aim of the project shown here is the artificial intelligence (AI)-supported automated object recognition and CAD model generation from laser scan data. The medium-term goal is the derivation of objects according to building information modeling (BIM).

CONTACT
Roland Degelmann
roland.degelmann@stmb.bayern.de

SOFTWARE
ArcGIS Pro 2.5

DATA SOURCES
Bayerische Staatsbauverwaltung; cloud-vermessung; supper&supper; Bayerische Vermessungsverwaltung

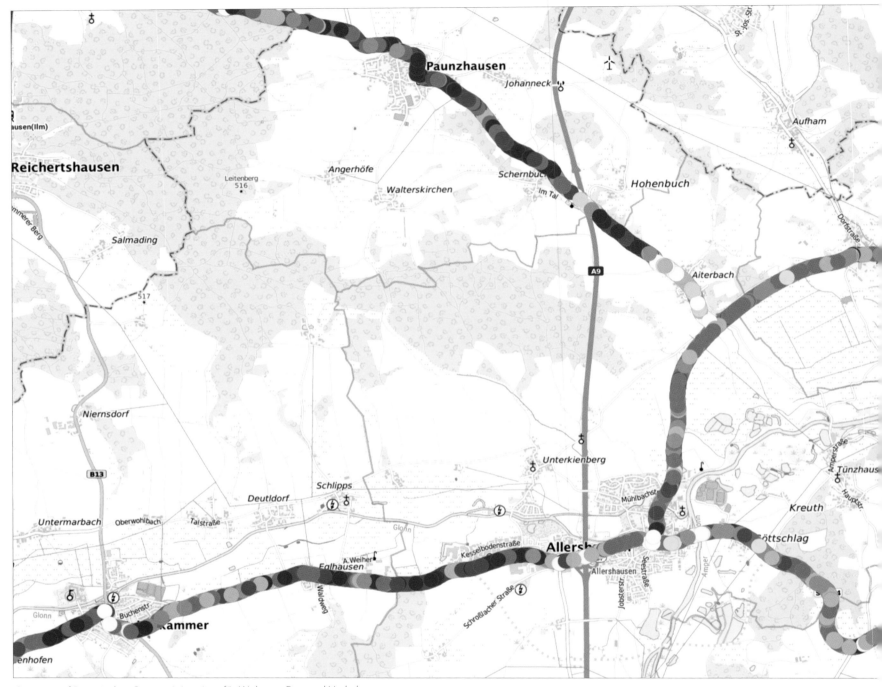

Courtesy of Bayerisches Staatsministerium für Wohnen, Bau und Verkehr.

TRAFFIC SPEED DEFLECTOMETER (TSD)

Bayerisches
Staatsministerium für
Wohnen, Bau und Verkehr
Munich, Germany
By Roland Degelmann

The assessment of structural conditions plays a crucial role in economic pavement maintenance management. The existing data of the monitoring and evaluation of pavement conditions provides a useful basis but is limited to the evaluation of surface conditions. The traffic speed deflectometer (TSD) allows for bearing capacity measurements in flowing traffic. In 2019, the entire Bavarian state road network was recorded using this technology.

124

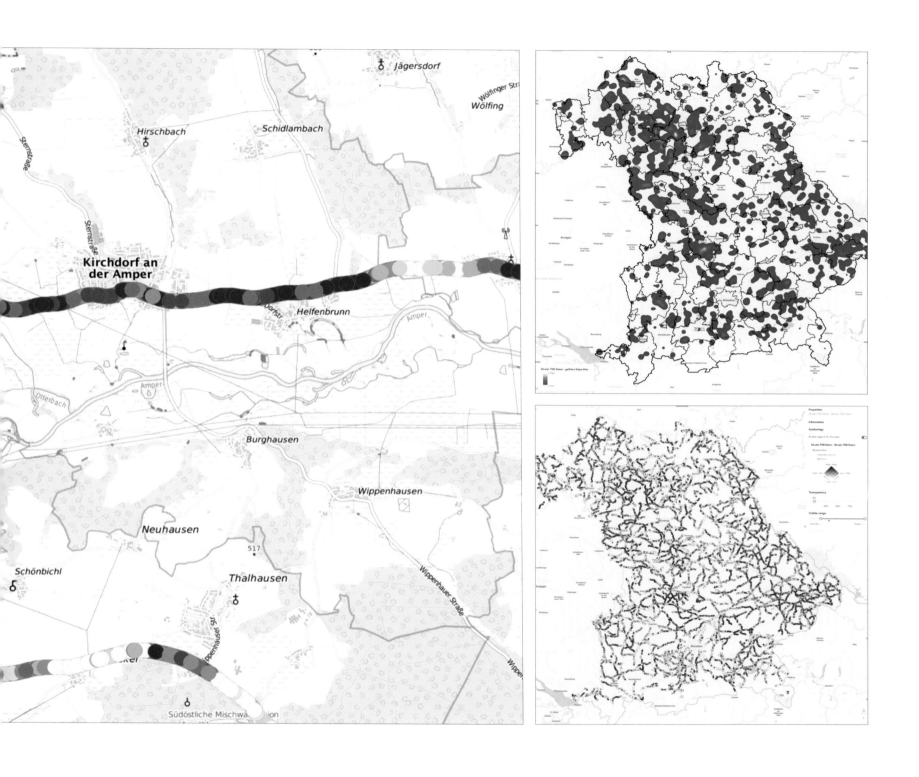

CONTACT
Roland Degelmann
roland.degelmann@stmb.bayern.de

SOFTWARE
ArcGIS Online

DATA SOURCES
TSD Data; Bayerische
Vermessungsverwaltung

MICHIGAN CENSUS 2020 RESPONSE RATES DASHBOARD

University of Michigan–Flint
Flint, Michigan, USA
By Troy Rosencrants

CONTACT
Troy Rosencrants
rosencra@umich.edu

SOFTWARE
ArcGIS Enterprise 10.7.1

DATA SOURCES
US Census Bureau

Through support from the Community Foundation of Greater Flint, the GIS Center at the University of Michigan–Flint created a series of web mapping applications to assist in Census 2020 outreach for Genesee County. One of the mapping applications created was an operations dashboard showing the self-response rates for Census 2020 by city and township for the state of Michigan.

The Michigan Census 2020 Response Rates Dashboard provided an avenue for community members to view how their community was responding to the census compared with other communities. The ability to provide data in lists, in addition to the map, allowed for the quick dissemination of information to assist in getting everyone counted in the 2020 census. The dashboard was created in April 2020 and has been updated daily as new rates were published by the US Census Bureau.

The map shows a layer of combined city and township boundaries symbolized by the cumulative response rates as of the most recent date with a county boundary layer overlaid. Each city or township can be selected to show a pop-up of information that shows the date, cumulative overall and internet response rates, and daily overall and internet response rates. In addition, the highest and lowest response rates by city/township and by county are provided as lists, which when selected highlight the location on the map.

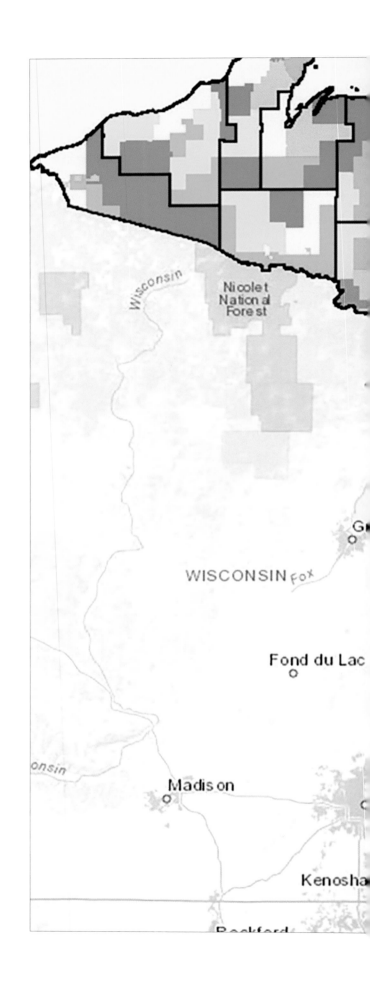

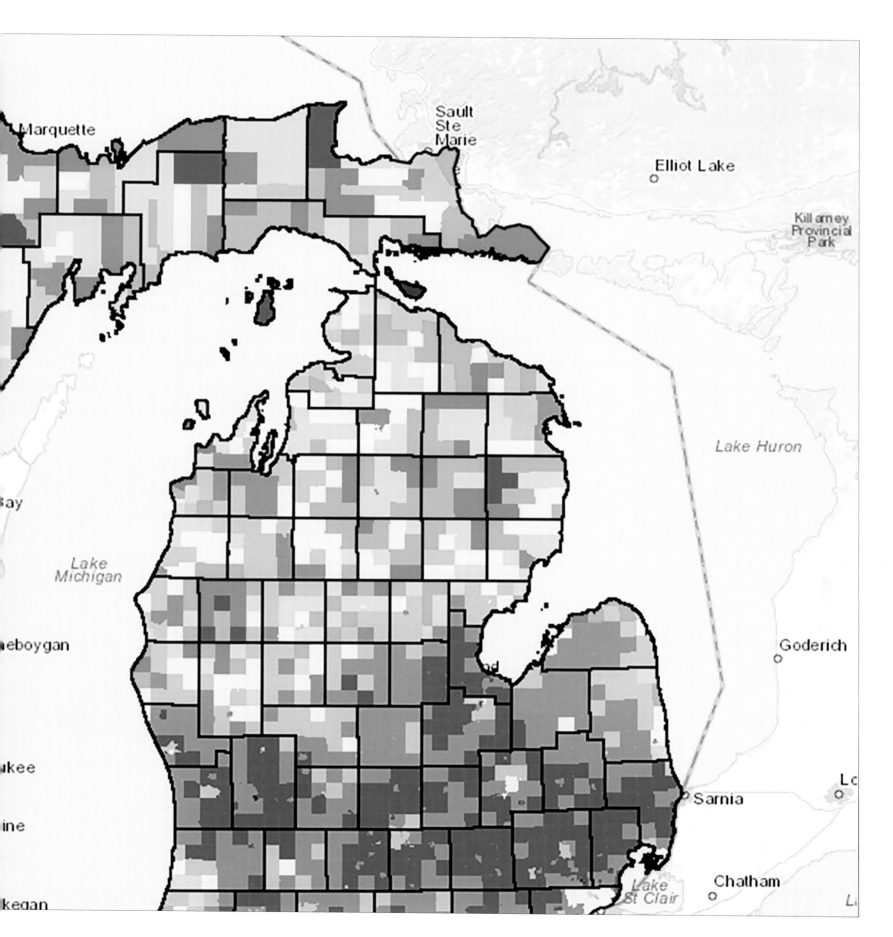

140 YEARS OF HUMAN IMPACT IN THE SOFIA VALLEY

Space Research and Technology Institute,
Bulgarian Academy of Sciences
Sofia, Bulgaria
By Stefan Stamenov and Vanya Stamenova

These maps show the increasing anthropogenic impact on the Sofia Valley in Bulgaria over the past 140 years. The Sofia city region is a good example of the growing human impact on nature during the last century. By the end of the 19th century, Sofia was a tiny town with only 12,000 citizens, and the Sofia Valley area was almost untouched by industrialization and modern infrastructure. Today, the city is home to 1.2 million citizens and is the most urbanized area in Bulgaria.

The map of the land cover situation in the Sofia Valley at the end of the 19th century shows fragments from one of the area's first topographic maps, dating from the 19th century, along with the boundaries of the Sofia Valley and the modern city of Sofia. The map of Sofia city's territorial expansion reveals the increase in area of the city from the end of the 19th century to the present and details the boundaries of the city in different years. A map of the land cover of Sofia in 1879 is also shown. The city's plans are georeferenced and vectorized within the project's GIS of the old Sofia.

Courtesy of Dr. Stefan Stamenov and Dr. Vanya Stamenova.

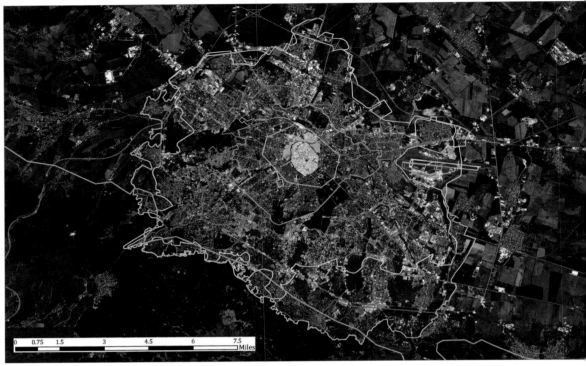

These maps represent the increasing of the anthropogenic impact on the territory of Sofia Valley over the last 140 years.

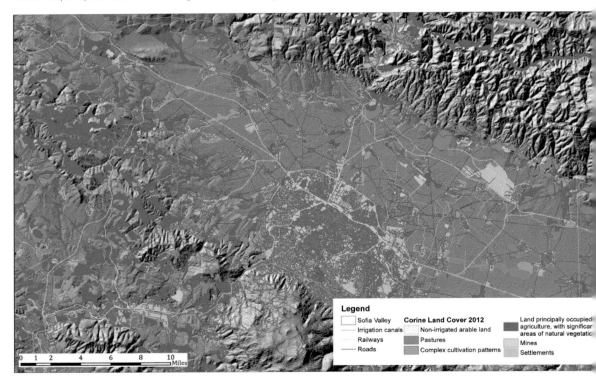

Legend

Sofia Valley | **Corine Land Cover 2012** | Land principally occupied agriculture, with significan areas of natural vegetatic
Irrigation canals | Non-irrigated arable land |
Railways | Pastures | Mines
Roads | Complex cultivation patterns | Settlements

	Sofia city 1879
Valley	
2009	Vegetable gardens
1982	Meadow
1961	Neighbourhood
1936	Fortification
1928	Streets
1892	Rivers

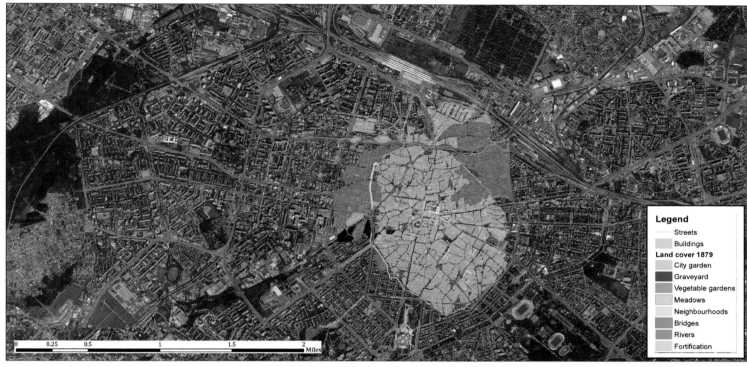

Legend
- Streets
- Buildings

Land cover 1879
- City garden
- Graveyard
- Vegetable gardens
- Meadows
- Neighbourhoods
- Bridges
- Rivers
- Fortification

Land cover of Sofia, representing the situation of the city of Sofia that corresponds to 1879.

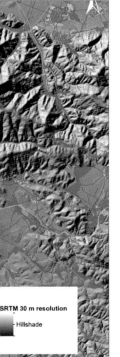

SRTM 30 m resolution

Hillshade

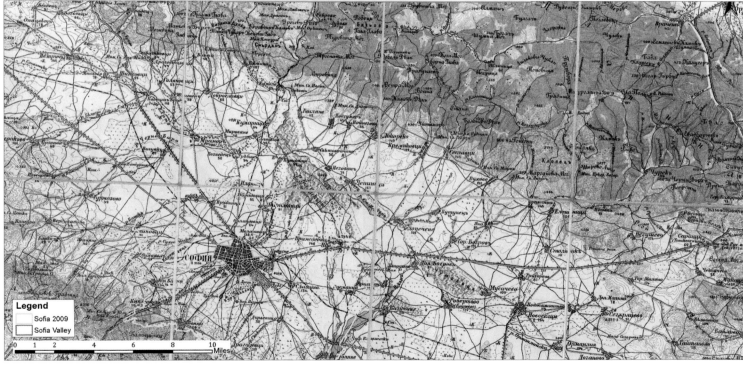

Legend
- Sofia 2009
- Sofia Valley

The map shows one of the first topographic maps from the end of the 19th century with the boundaries of Sofia Valley and the modern city of Sofia in 2009.

CONTACT
Stefan Stamenov
stamenovstefan@yahoo.bg

SOFTWARE
ArcMap, ArcGIS Desktop 10.7,
ECP grant to SCGIS Chapter
Bulgaria

DATA SOURCES
USGS; Copernicus Land Monitoring Service; JICA Database for Bulgaria,
2006; National Library St. St. Cyril and Methodius; DigitalGlobe Foundation;
ESA Copernicus Sentinel Data – Copernicus Open Access Hub

THE MEKONG INFRASTRUCTURE TRACKER

Blue Raster
Arlington, Virginia, USA
By Blue Raster, Stimson Center

CONTACT
Michael Lippmann
mlippmann@blueraster.com

SOFTWARE
ArcGIS Online, ArcGIS API for JavaScript

DATA SOURCES
The Stimson Center

Home to more than 300 million people and some of the most endangered wildlife on the planet, the Greater Mekong region in Southeast Asia is the second most naturally diverse place in the world, second only to the Amazon. This region is also home to the Mekong River, which sustains not only much of the wildlife in the area but also the people. The majority of the population in this region depends on the river and its surrounding wetlands to support their way of life. In turn, this also means that this population and the vast biodiversity in the region are vulnerable to the economic and environmental effects of growing infrastructure.

The Stimson Center works to address the environmental and social impacts along the Mekong River to improve food security, stability, and cross-country relations. To help analyze these issues, Blue Raster collaborated with the Stimson Center to create the Mekong Infrastructure Tracker, an interactive web app to explore the infrastructure boom and its impacts in the Mekong region. With funding provided by USAID, the Mekong Infrastructure Tracker was developed with support from the USAID Mekong Safeguards activity led by the Asia Foundation.

The Mekong Infrastructure Tracker web app uses ArcGIS Online and the ArcGIS API for JavaScript to provide users with data transparency for analyzing the type and scale of different infrastructure projects in the region in relation to socioeconomic and environmental factors.

Courtesy of Blue Raster.

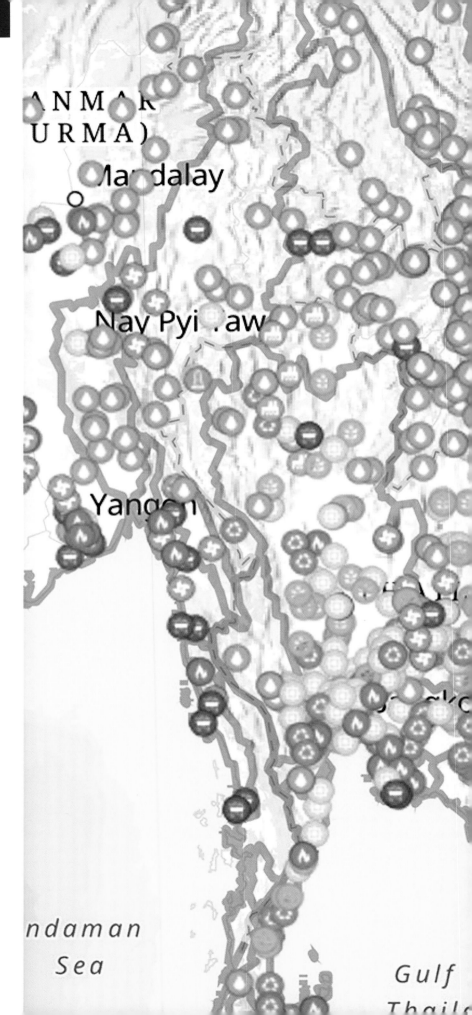

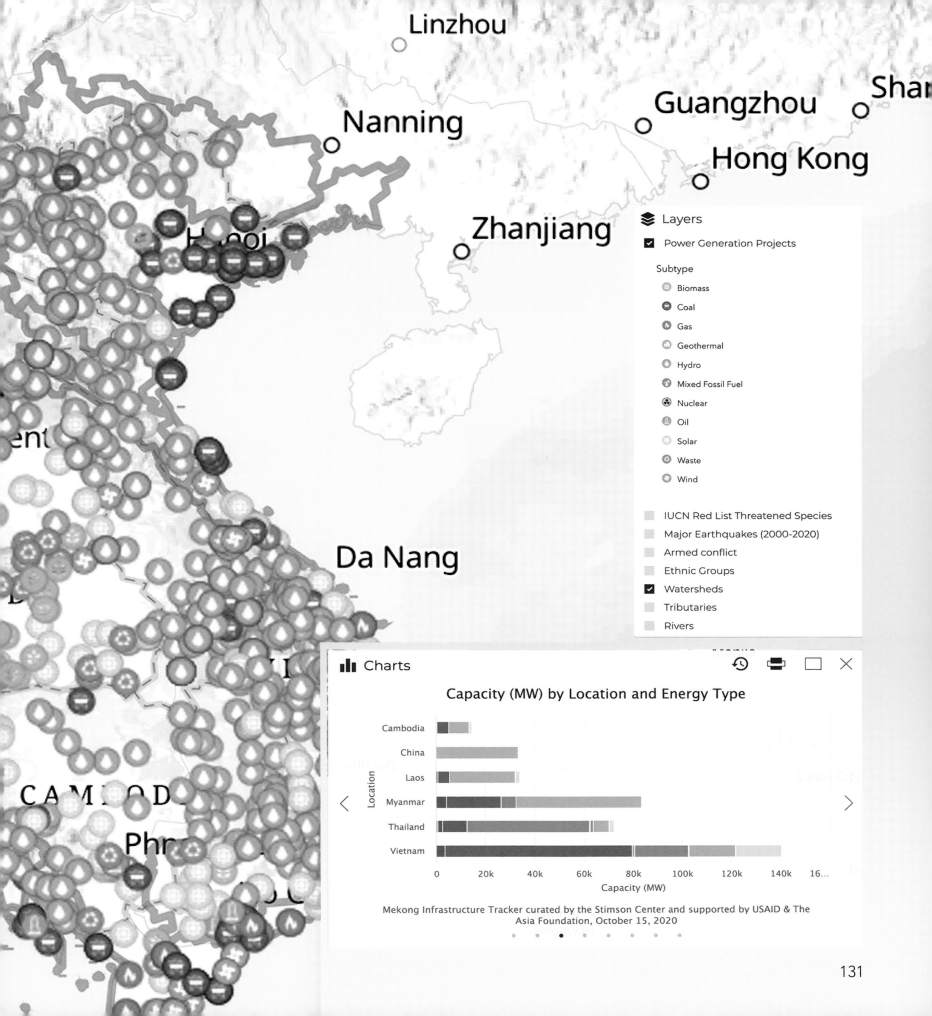

Linzhou

Shar

Guangzhou

Nanning

Hong Kong

Zhanjiang

Layers

☑ Power Generation Projects

Subtype

⊙ Biomass
⊖ Coal
◈ Gas
◎ Geothermal
◉ Hydro
⊕ Mixed Fossil Fuel
⊛ Nuclear
⊙ Oil
◯ Solar
⊗ Waste
⊕ Wind

☐ IUCN Red List Threatened Species
☐ Major Earthquakes (2000-2020)
☐ Armed conflict
☐ Ethnic Groups
☑ Watersheds
☐ Tributaries
☐ Rivers

Da Nang

CAMBOD

Phr

▉ Charts   ⟲ 🖶 ☐ ✕

Capacity (MW) by Location and Energy Type

Cambodia
China
Laos
Myanmar
Thailand
Vietnam

0   20k   40k   60k   80k   100k 120k 140k 16...
Capacity (MW)

Mekong Infrastructure Tracker curated by the Stimson Center and supported by USAID & The Asia Foundation, October 15, 2020

131

Courtesy of Houseal Lavigne Associates.

PLANNING, VISUALIZING, AND EXPERIENCING A TOWN CENTER

Houseal Lavigne
Associates
Chicago, Illinois, USA
By Devin Lavigne, Nik
Davis, and Daniel Tse

For Morrisville, North Carolina, indecision had become a stalemate for its town center project. Despite community support to create a downtown area, local officials were apprehensive about making lasting decisions for the development. What the community needed was more than just a plan, but a way for officials to experience the town center and understand how

design standards, zoning regulations, and site plans would actually impact the development of the site. The town hired Houseal Lavigne Associates to solve the stalemate by planning and designing a 3D visualization of the town center site. Moving beyond pictures and graphics, Houseal Lavigne designed a fully immersive 3D model that allowed officials to explore multiple scenarios and walk through the

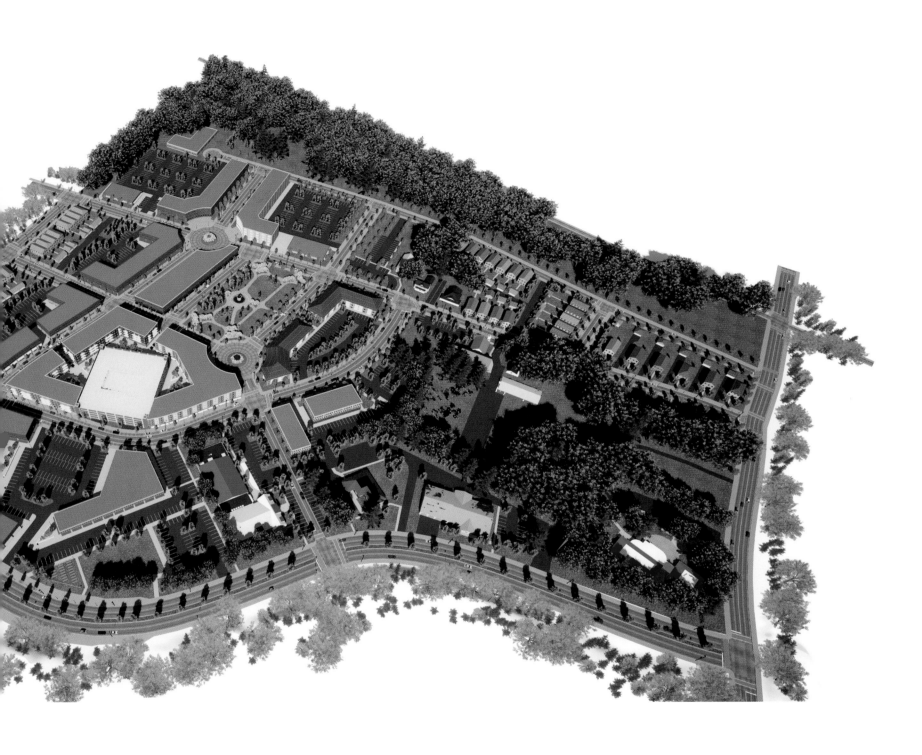

site from a pedestrian's perspective. This interactive map describes the challenges Morrisville faced and the process used to develop the town center visualizations that ultimately provided local officials with the perspective necessary to take action for their town center.

CONTACT
Devin Lavigne
dlavigne@hlplanning.com

SOFTWARE
ArcGIS Desktop, ArcGIS CityEngine, ArcGIS Pro, ArcGIS Online, Unreal Engine, SketchUp, Photoshop

DATA SOURCES
Living Atlas,
Town of Morrisville

GEOGRAPHY OF A STADIUM

Houseal Lavigne Associates
Chicago, Illinois, USA
By Devin Lavigne, Sujan Shrestha, and Brian Sims

CONTACT
Devin Lavigne
dlavigne@hlplanning.com

SOFTWARE
ArcGIS Desktop, ArcGIS Pro, Esri Story Maps,
NearMap, Photoshop

DATA SOURCES
Miller Coors, Ticketmaster, Pittsburgh Steelers, NearMap

Heinz Field is a football stadium in the North Shore neighborhood of Pittsburgh, Pennsylvania. It serves as the home of the Pittsburgh Steelers of the National Football League. Attendance for the 68,400-seat stadium has sold out for every Steelers home game, a streak that goes back to 1972. Intrigued by data provided by the MillerCoors Co., Houseal Lavigne Associates used the power of ArcGIS to analyze the distribution of beer and other amenities throughout the stadium. A routing network was created with rows, aisles, and concourses, connecting seats to concession areas and bathrooms. Costs were assigned to stairs and rows to deter them as walkable routes, and ultimately the stadium was mapped to determine the accessibility to beer, by type (bottles or package), and by brand.

Houseal Lavigne incorporated this data into an interactive map app that can help fans navigate the stadium as well as assist beer companies and stadium operators with marketing, inventory, and ticket value decisions and identify locations where beer sales can be expanded. In addition, the project sets the precedent for analyzing the design of stadiums to determine ways that accessibility to key amenities and destinations can be improved in the future.

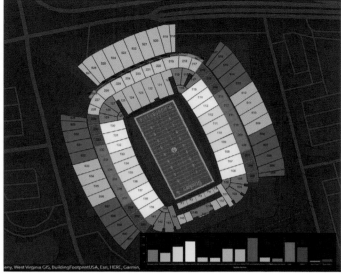

Heinz Field Explorer

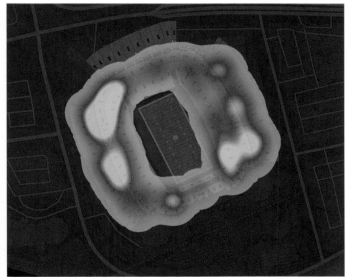

Location of Products

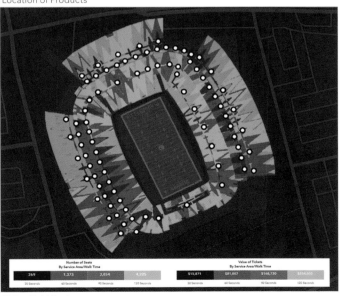

Heinz Field Explorer

135

SINGLE-FAMILY PLATTING ACTIVITY, 2019

Planning & Engineering Department,
City Of Houston
Houston, Texas, USA
By Sona Zechariah

CONTACT
Sona Zechariah
Sona.Sunny@houstontx.gov

SOFTWARE
ArcGIS Pro 2.4.3

DATA SOURCES
COHGIS

Houston, Texas, is the fourth largest city in the United States. It is a dynamic, growing city with an area of 671 square miles. This map shows the 2019 Single-Family Residential Platting Activity within Houston's city limits and its extraterritorial jurisdiction (ETJ). The data was collected using the City of Houston's Plat Tracker application, which allows land planners, civil engineers, and surveyors to submit subdivision plats for review and presentation to the Houston Planning Commission. Plats are checked for the proper subdivision of land, adequate street or right-of-way, building lines, and compliance with Chapter 42 (the city's land development ordinance). This map is used to visualize where the development and growth is occuring. The size of the dots represents the number of lots per plat.

Courtesy of City of Houston.

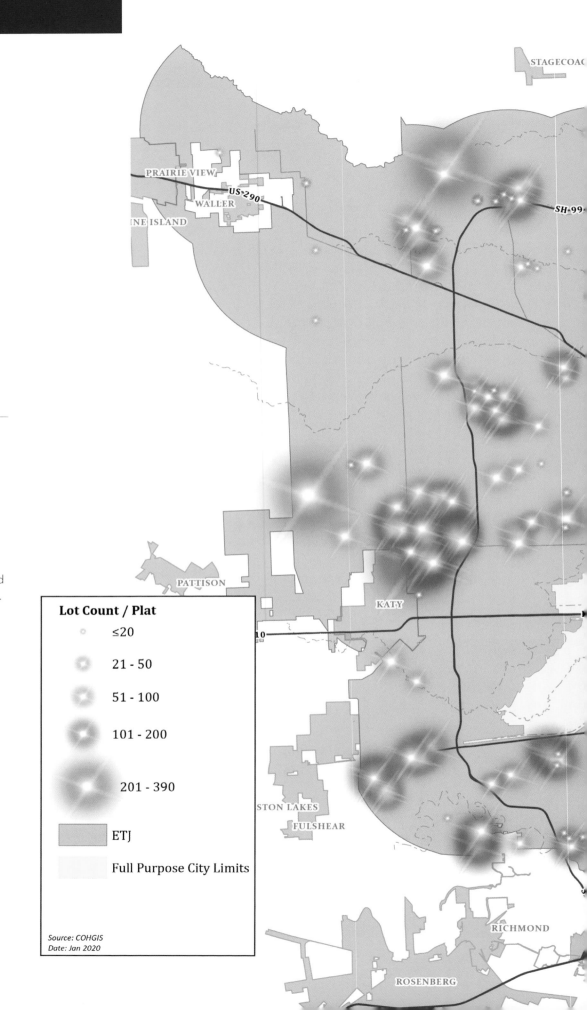

Lot Count / Plat

≤20

21 - 50

51 - 100

101 - 200

201 - 390

ETJ

Full Purpose City Limits

Source: COHGIS
Date: Jan 2020

Courtesy of SymGEO.

VIEWSHED ANALYSIS OF A PROPOSED STRUCTURE

SymGEO
Gaithersburg, Maryland, USA
By Kevin McMaster

The visual impact of new construction may require significant community outreach and education during the permitting and preconstruction process. Esri GIS tools, including 3D basemap production using high-accuracy lidar data, greatly facilitate the viewshed analysis of proposed structures and scenarios. For the

Kentlands neighborhood in Gaithersburg, Maryland, SymGEO built a digital twin to show the 3D digital modeling and analytic capabilities of ArcGIS Pro and ArcGIS CityEngine. Realistic building textures and intuitive symbology allowed planners to see which areas would be able to see the new building (in green), or

which areas would not have a clear sightline of the proposed structure (in red). The results of this modeling could be used to inform planners where residents may need additional outreach efforts during the planning phase. Alternative scenarios can be quickly incorporated and viewsheds rendered to compare proposed construction elements.

CONTACT
Kevin McMaster
kevin.mcmaster@symgeo.com

SOFTWARE
ArcGIS Pro, ArcGIS CityEngine

DATA SOURCES
The Maryland-National Capital Park and Planning Commission

A GAME OF SHADOWS

University of San Diego, San Diego, California, USA
By Soydan Alihan Polat

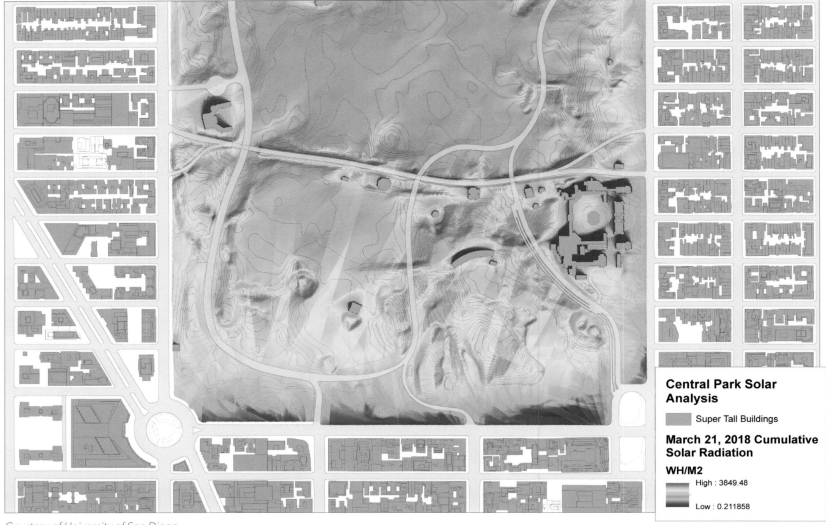

Central Park Solar Analysis

■ Super Tall Buildings

March 21, 2018 Cumulative Solar Radiation

WH/M2

High : 3849.48

Low : 0.211858

Courtesy of University of San Diego.

This project is about the Manhattan skyline and its effect on Central Park. The general concern about the height of tall buildings is shadows. But how do we quantify shadows? I performed many shadow analyses, showing the cumulative effects of shadows with hourly increments. The important thing is not the shadow or shade but the lack of sunlight. We always quantify the effect of building shadows, but we should, in

fact, quantify the loss of sunlight. And sunlight—in raw terms—is solar radiation.

Through many iterations and a few weeks of computation time, I performed a solar radiation analysis of the entire Central Park for a single day, March 21—vernal equinox (the first day of spring). It's one of the four days city environmental quality review (CEQR) mandates for shadow analysis.

Three "supertalls" that are under construction with recent opening dates are: 111 West 57th Street (Steinway Tower), 220 Central Park South, and Central Park Tower (Nordstrom Tower). The analysis was performed twice, with and without these three towers. The difference between the two sets of data analysis gave us the net loss of solar radiation for Central Park. This map summarizes the loss of solar radiation on a 50-by-50-foot grid. The units for measuring

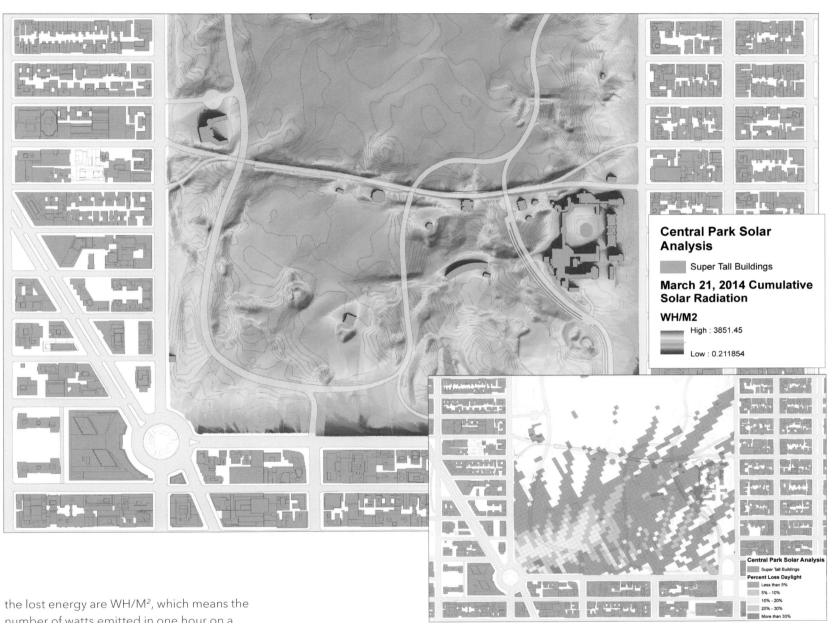

Central Park Solar Analysis

⬛ Super Tall Buildings

March 21, 2014 Cumulative Solar Radiation

WH/M2

High : 3851.45

Low : 0.211854

Central Park Solar Analysis

⬛ Super Tall Buildings

Percent Loss Daylight

Less than 5%

5% - 10%

10% - 20%

20% - 30%

More than 30%

the lost energy are WH/M², which means the number of watts emitted in one hour on a square meter surface.

The utility company measures the energy usage by kilowatt hours. Each 50-by-50-foot cell area is converted to square meters and multiplied by the average energy loss (in WH/M²) with each cell's square meter area to find the total average energy loss for each cell (WH).

CONTACT
Soydan Alihan Polat
spolat@sandiego.edu

SOFTWARE
ArcGIS Desktop with 3D Analyst and Spatial Analyst

DATA SOURCES
NYC 3D buildings (multipatch), 2' contours

CAPITAL CRITERIA–BLOCK BY BLOCK

Unified Government of Wyandotte County, Kansas
Kansas City, Kansas, USA
By Robert Anderson

CONTACT

Robert Anderson
randerson@wycokck.org

SOFTWARE

ArcGIS Pro

DATA SOURCES

UG Wyco Data, UG Public Works, UG GeoSpatial Services, UG
Finance, UG Board of Commissioners, Board of Public Utilities (Kansas
City, KS), Peach Tree Data

This map depicts two opposing forces in local government by block:
a landscape of commission priorities based on a weighted capital
criterion (ascending in green) and the costs to replace or maintain
public infrastructure (descending in red). It is represented in a block-
by-block format that links parcel data with the value and costs of
public infrastructure in the right-of-way. In building a system to rank
capital improvement projects, the Unified Government of Wyandotte
County/Kansas City, Kansas (UG) Asset Management team computed
an enriched parcel layer to inform analysis buffers of unfunded capital
projects using the commission's weighted priorities. The parcel
layer contains a real-estate tax roll, estimated sales tax revenues,
population data per parcel, community sites, project coordination
potential, asset condition analysis, average daily traffic, planning
data, and more. The cost of infrastructure contains the estimated
maintenance and replacement costs on multiple asset classes (roads,
sidewalks, curbs, streetlights, and so on).

Both layers were summarized within their blocks to compute a
relationship between priorities and costs per block. A key commission
goal is measuring the return on investment of projects, and this data
estimates the years of revenue needed to fund individual asset classes
for each capital project. This visualization is also an interactive 3D
map that offers a discussion point for the elected body to understand
the landscape that its weighted capital criteria input creates in the
scoring process. Showing infrastructure costs per block renders
visible the funding challenges at a uniquely human scale.

*Data designed and prepared by UG Public Works Asset Management & the UG
Wyco Data Team: Robert Anderson, Maddie McNerthney, Hannah Mosiniak,
and Jud Knapp.*

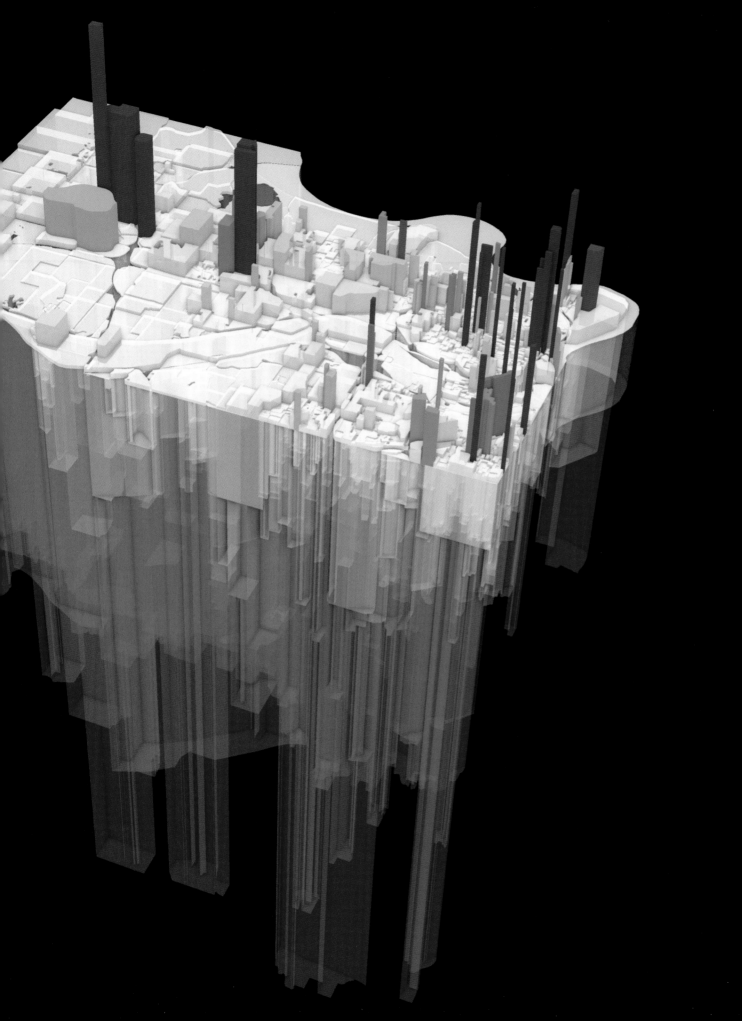

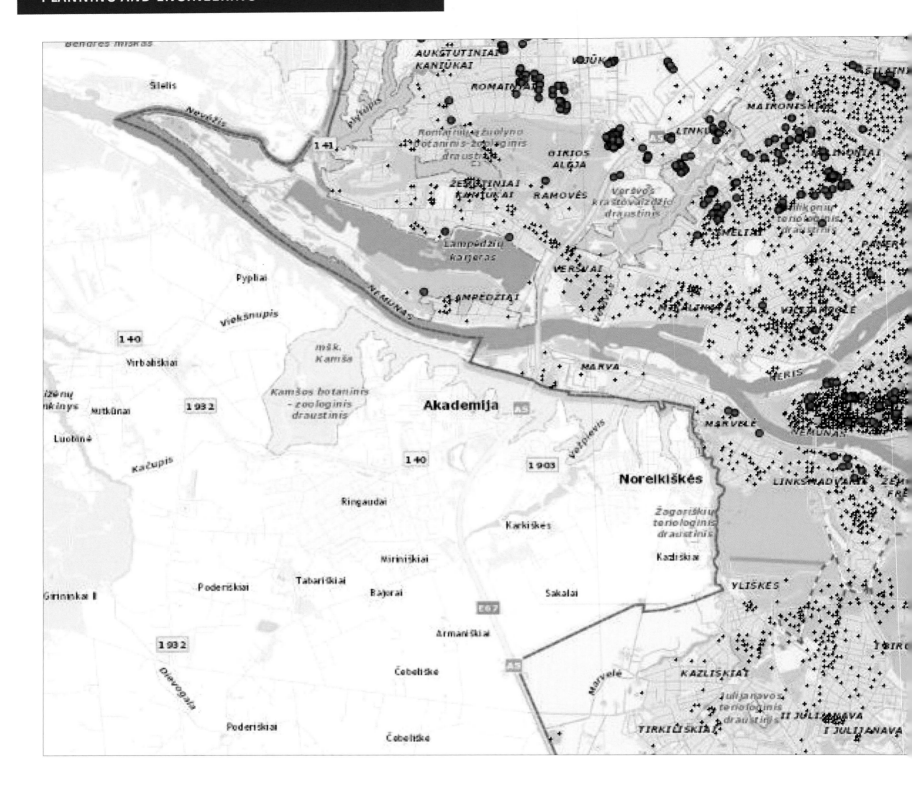

ABANDONED REAL PROPERTY IN KAUNAS CITY, LITHUANIA

Kauno planas, ME
Kaunas, Kauno, Lithuania
By Arunas Sipavicius

A map of abandoned real property in Kaunas City, Lithuania. The property objects have been registered and monitored using ArcGIS Survey123.

The municipality staff uses the interactive map to search for property owners, initiate fines, and track changes in property maintenance.

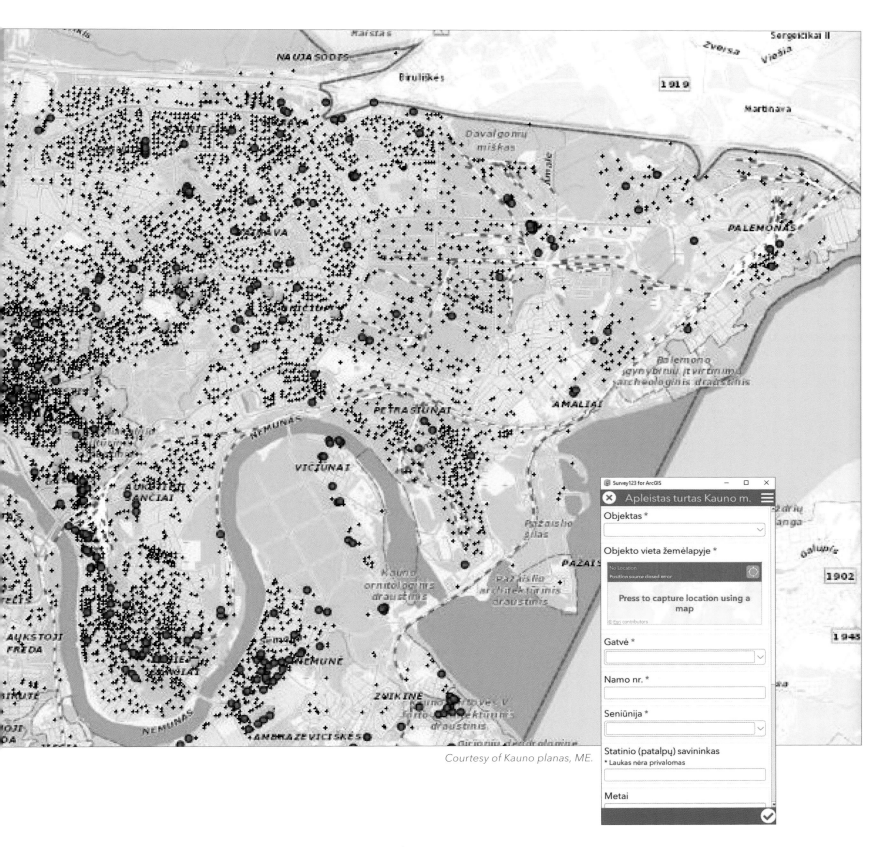

Courtesy of Kauno planas, ME.

CONTACT
Arunas Sipavicius
A.Sipavicius@kaunoplanas.lt

SOFTWARE
ArcGIS Enterprise

DATA SOURCES
Kauno planas

145

RAK MUNICIPALITY LAND USE APPLICATION, RAK GEOMAP

RAK Municipality
Ras Al Khaimah, United Arab Emirates
By Aaesha Saif Shehhi

CONTACT
Aaesha Saif Shehhi
aaesha.shehhi@mun.rak.ae

SOFTWARE
ArcGIS Enterprise 10.6.1, ArcGIS Web AppBuilder,
ArcGIS API for JavaScript

DATA SOURCES
Town planning master plan and business information

Municipal land use maps are a reliable source of authorized information from the town planning department of RAK municipality, where updates are performed in near real time, as and when the transaction happens. Esri Spatial Database Engine (SDE) hosted, and ArcGIS Enterprise-based map services enabled easy, secure land use map distribution. Classification and cartography provided ease of recognition, ease of validation with an underlying basemap, and aerial imagery.

This land use map is operational for various disciplines and users, for a variety of daily tasks, including engineering, planning, surveying, and land registry. These users interactively identify, query, filter, map, print, locate, and apply advanced queries using spatial selection functionality. The application enables maps to be distributed within an organization with greater efficiency. In addition, integration with SAP data allows users to view real estate and building information using REST technology.

Courtesy of RAK Municipality.

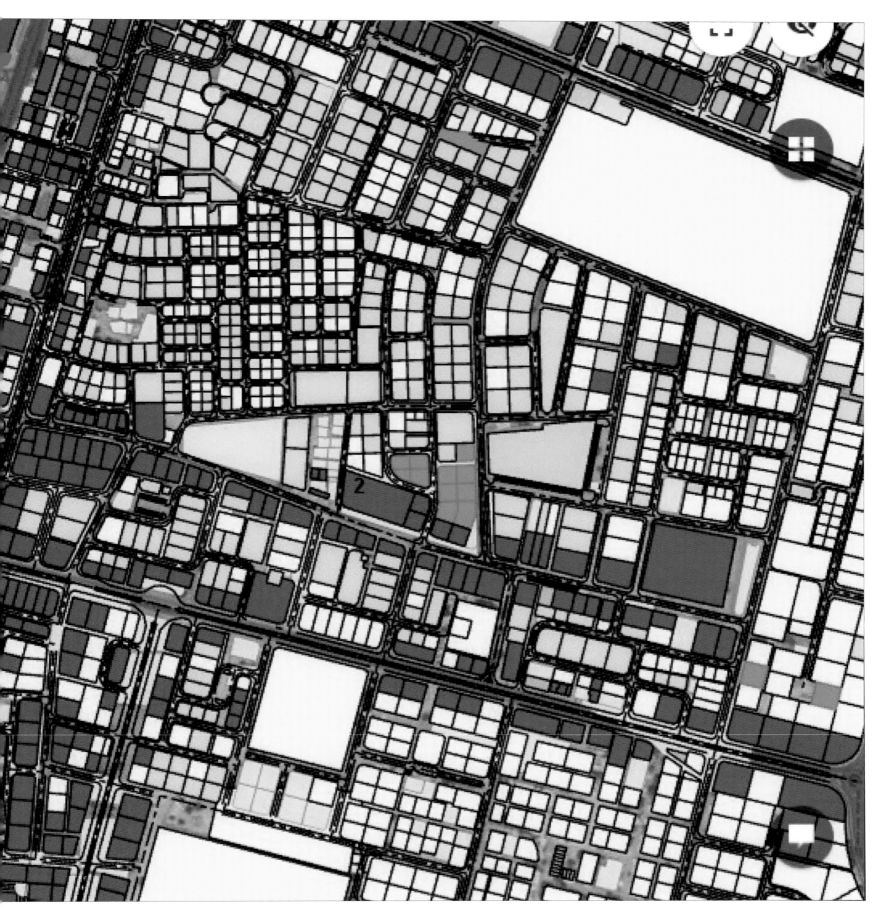

147

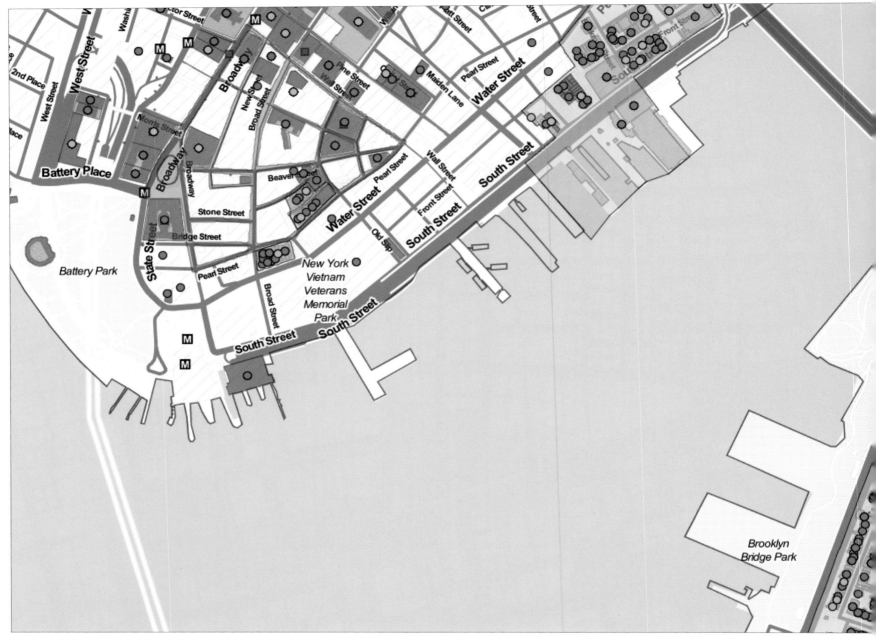

Courtesy of New York City Landmarks Preservation Commission.

PERMIT APPLICATION FINDER

New York City Landmarks Preservation Commission New York, New York, USA By New York City Landmarks Preservation Commission

The New York City Landmarks Preservation Commission (LPC) Permit Application Finder map is a publicly accessible, searchable database with information on all applications and permits the agency issues for work on designated landmark buildings and sites across the five boroughs. The web map includes detailed information on all applications filed with the LPC since January 1, 2016,

including received date, work type, applicant information, issue date, and a copy of the permit. Information on the map is updated daily as permits move through the agency from filing to issue.

To use the map, members of the public can search by address or docket number (the unique permit identification number), or they can click on designated

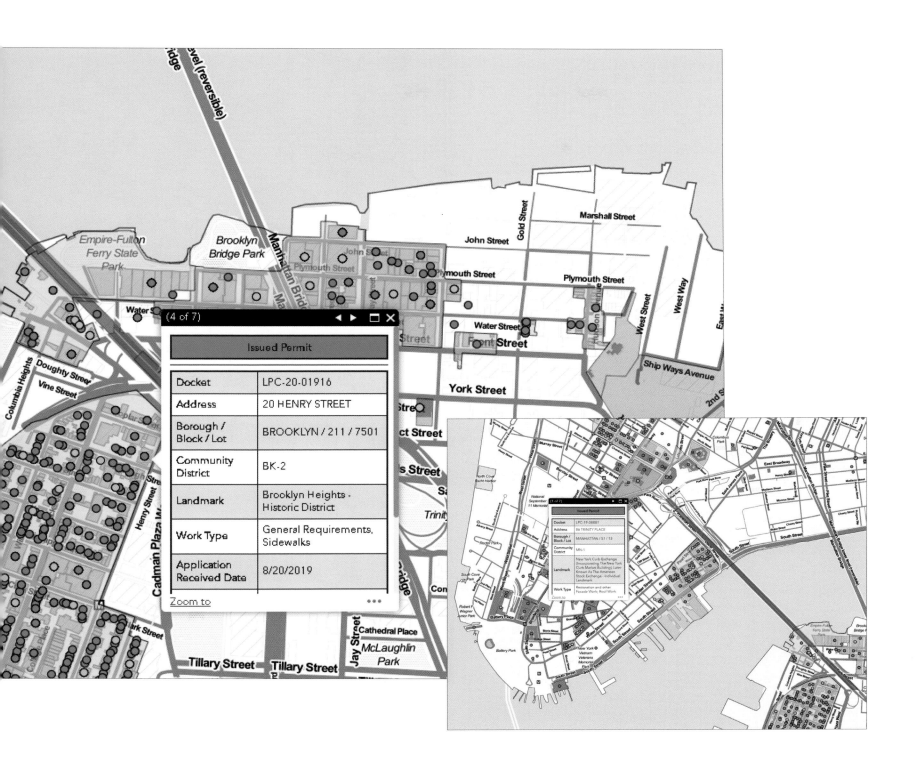

buildings and sites to see whether there are any applications for work at that location that require an LPC permit.

The Permit Application Finder increases transparency between the agency and the public, allowing anyone to access the approximately 14,000 applications the LPC processes in a typical year. Members of the public can see what work is being done on designated buildings throughout the city, and applicants can check on the status of their permit.

CONTACT
Timothy Frye
tfrye@lpc.nyc.gov

SOFTWARE
ArcGIS Online

DATA SOURCES
All landmark and permit data by New York City Landmarks Preservation Commission.

GEOAI TECHNIQUES IN PRE-ENUMERATION CASE STUDIES

Federal Competitiveness and Statistics Authority (FCSA)
Dubai, United Arab Emirates
By Marwa Elkabbany

CONTACT
Marwa Elkabbany
marwa.elkabbany@fcsa.gov.ae

SOFTWARE
ArcGIS Enterprise, ArcGIS Pro

DATA SOURCES
UAE Federal Competitiveness and Statistics Authority;
Federal Electricity & Water Authority

This map represents a prototype on how geospatial artificial intelligence (GeoAI) can be used in the census pre-enumeration, which combines extracted buildings from imagery with smart meter data to identify the land use and accommodation status using meter information.

The process started by segmenting and classifying satellite imagery to identify building features. These building features were then extracted and overlaid with existing residential electricity and water meters to identify buildings with and without meters.

Buildings without residential electricity and water features are considered new housing units, requiring surveying and inclusion in the delineation of enumeration areas.

Through image classification and segmentation, it was determined that 84 percent of the buildings extracted are served by residential meters in 2017.

TOTAL DENSITY

Buildings with
no meters

Buildings with
meters

151

IMPLEMENTING THE NEW HARVARD INTERACTIVE WEB MAP WITH ARCGIS INDOORS™

Harvard University, Cambridge, Massachusettes, USA
By Harvard Planning Office: Jim Nelson, Parvaneh Kossari, Giovanni Zambotti, and Michael Guarino

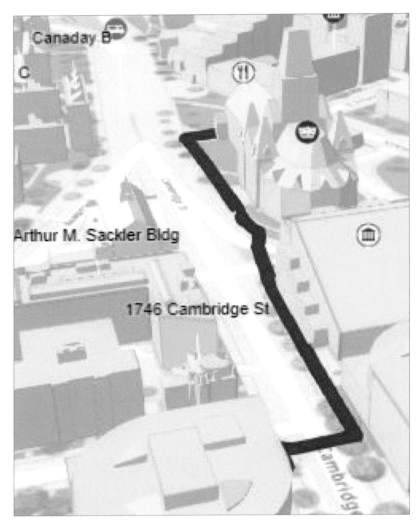

Courtesy of Harvard University.

The new public Harvard Interactive Map strives to be a user-friendly accessible source of campus information that is compatible with current browsers and devices. The map provides real-time information on events, university shuttles, and bike share availability. Walking directions for the shortest routes work with a pedestrian network that includes campus pathways and building interiors in three dimensions. Routes without mobility barriers can also be calculated. The application is scalable and capable of providing secure access to asset and space information.

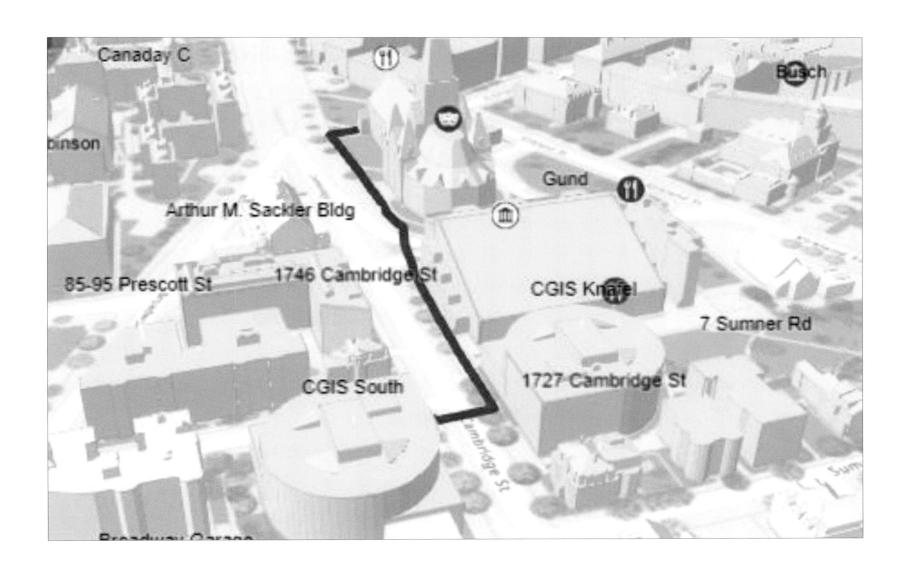

CONTACT
Parvaneh Kossari
parvaneh_kossari@harvard.edu

Jim Nelson
james_nelson@harvard.edu

SOFTWARE
ArcGIS Indoors, ArcGIS Pro

DATA SOURCES
Harvard University

BRNO CITY 3D MODEL

Brno City Municipality
Brno, Jihomoravsky, Czech Republic
By Anna Hradecka, Kristyna Zednickova, and Jan Fiser

CONTACT
Dana Glosova
glosova.dana@brno.cz

SOFTWARE
ArcGIS Pro, Terinos, Adobe InDesign

DATA SOURCES
Statutory City of Brno

The 3D model of the city of Brno, Czech Republic, is being built as a part of the Digital Map of Brno. It is based on a precise map known as the Map of Technical Utilization of the area. The map contents are combined with other asset management systems, including tree, road, and street furniture asset management systems, all in 2D. These systems have been extended with a 3D model of buildings in Level of Detail 1 (LOD1), which is continuously updated and integrated with up-to-date elements from other datasets. The aim of the map is to allow users to work in a digital environment with a complex and up-to-date 3D model of the area.

This model is used also in the Terinos web application, which allows for adding objects to the area's newest model. Experts can make decisions about planned urban development options or review newly proposed buildings, especially about their height parameters in 3D directly. The application can also be used for modeling greenery systems or other phenomena. In the future, the application will allow for the presentation of planned projects to the public.

The project objective is more efficient city administration and improved communication with the public.

Copyright 2020 Statutory City of Brno, Copyright 2020 T-MAPY.

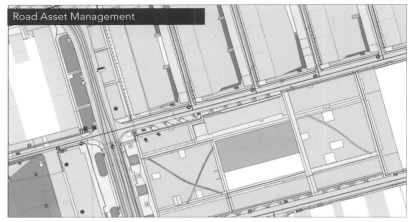

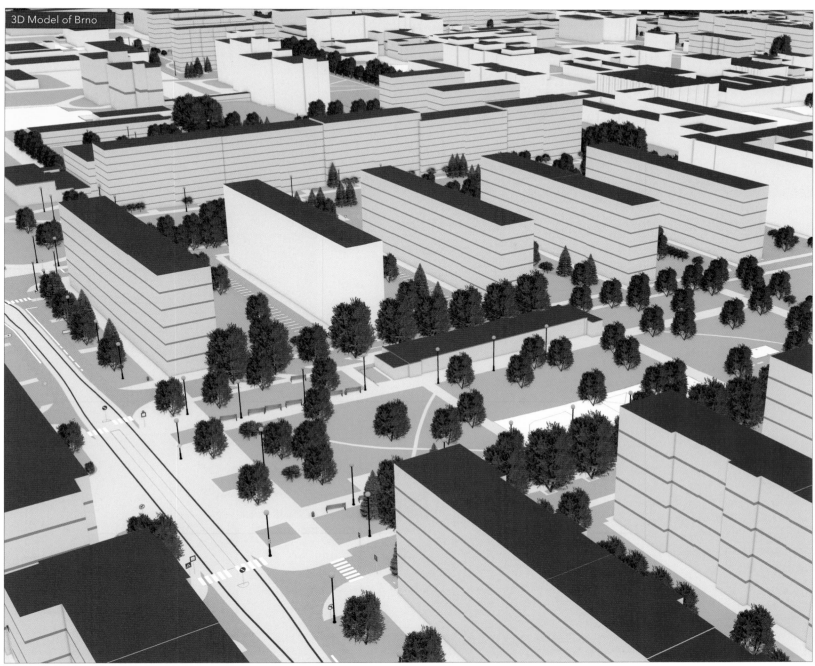

3D Model of Brno

Terinos Web Application

New Proposed Building

SCHOOLCRAFT COLLEGE CAMPUS FACILITIES WEB APP

Fishbeck
Grand Rapids, Michigan, USA
By Michelle Lazar and Caryn Ashbay

CONTACT
Michelle Lazar
mclazar@fishbeck.com

SOFTWARE
ArcGIS Desktop 10.7, ArcGIS Web AppBuilder

DATA SOURCES
Schoolcraft College, Livonia, Michigan

The Department of Construction and Design at Schoolcraft College is creating a mapping application to assist in facilities management of its interior spaces. The college maintains facilities at three locations, with 17 buildings located on its main campus in Livonia, Michigan. As is common in a campus environment, building plans, room accommodations, material finishes, and equipment specifications were historically stored in multiple files and databases. The goal of the project is to store this information in a single platform that also provides maintenance staff, space schedulers, and administrators with a tool to view campus assets from a standardized database using desktop and mobile apps. The program will assist staff in reporting on the current status of campus facilities and will be used for repurposing space and future development. The interior space features were converted from AutoCAD drawing files and include information to track the Facilities Inventory and Classification Manual (FICM) code and interior space finishes such as wall and floor coverings, lighting, and plumbing fixtures. The Americans with Disabilities Act (ADA) requirements for each room are also stored, allowing users to quickly verify room suitability when scheduling classes. The program has grown to include specifications for mechanical equipment, electric panel boards, and wireless access points, which may eventually be used to develop indoor navigation.

Courtesy of Fishbeck.

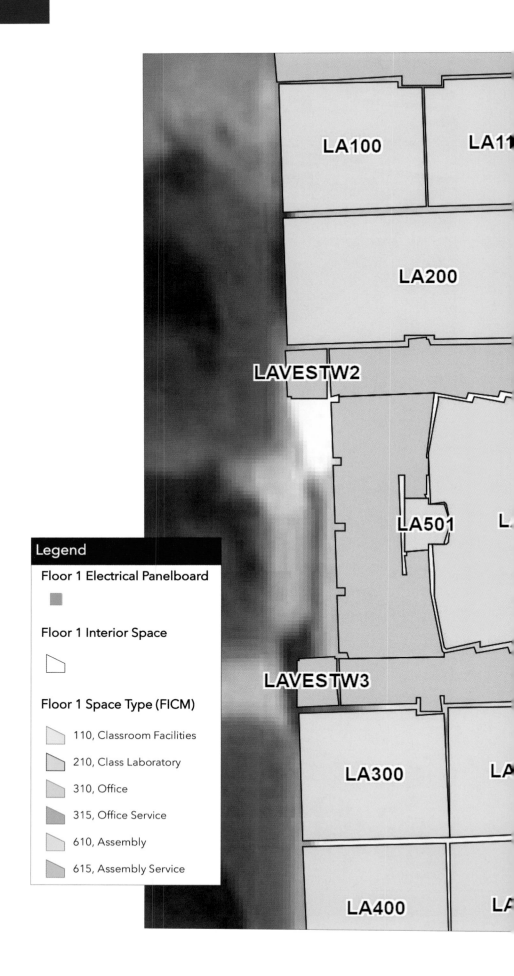

FUTURE LAND USE MAP

City of Round Rock
Round Rock, Texas, USA
By Kim Jones

CONTACT
Kim Jones
kjones@roundrocktexas.gov

SOFTWARE
ArcGIS Desktop 10.7

DATA SOURCES
City of Round Rock

The City of Round Rock strives to be the best medium-sized city in the country for people to live, work, and recreate and continues to achieve that goal through thoughtful planning in the face of relentless growth. The city's Planning and Development Services Department develops a new, comprehensive plan every 10 years to achieve that goal. We analyze past and recent growth, engage with landowners and the community, and formulate the plan to manage the city's growth in a way that benefits residents and businesses alike. The Future Land Use map was created as an exhibit for the Round Rock 2030 comprehensive plan.

Round Rock 2030 was adopted by the city council in June 2020 after two years of extensive public input and discussion. It now serves as the policy guide and framework for land use decisions for the next decade. The plan highlights how the city has changed over the past decade and creates a vision for what the community wants Round Rock to look like in 2030, according to its vision statement: "Round Rock is a safe, desirable, family-oriented community that balances progress and prosperity with its history, by prioritizing quality of life, mobility, economic development, and thoughtful land use planning."

Courtesy of City of Round Rock.

TRANSPORTATION (2017)

EGIONAL ATTRACTION

BLIC FACILITIES

NING

TY LIMITS

FREEWAY/TOLLWAY

FM/RM/STATE

ARTERIAL

CR 114/CHANDLER RD

CR 111

FM 1460

UNIVERSITY BLVD

CR 112

CR 110

LIMMER LOOP

CHISHOLM TRAIL RD

ARTERIAL L

SUNRISE RD

COLLEGE PARK DR

N. AW GRIMES BLVD

N RED BUD LN

CR 113

E OLD SETTLERS BLVD

RS BLVD

N KENNEY FORT BLVD

130

TIGER TRL

BOWMAN RD

N MAYS ST

79

S KENNEY FORT BLVD

RED BUD LN

ROUND ROCK AVE

DEEP WOOD DR

S MAYS ST

S AW GRIMES BLVD

FOREST CREEK

DOUBLE CREEK DR

GATTIS SCHOOL RD

NEIL RD

GREENLAWN BLVD

CR 170

SCHULTZ LN

45

CR 172

TOURISM AND RECREATION

GRAVE SMOKY MOUNTAINS NATIONAL PARK

Florida Department of Environmental Protection
Tallahassee, Florida, USA
By Aaron Koelker

CONTACT

Aaron Koelker
Aaron.Koelker@FloridaDEP.gov

SOFTWARE

ArcGIS Pro 2.5

DATA SOURCES

US National Park Service; US Board on Geographic Names

This map was a personal passion project. While most people visit Great Smoky Mountains National Park to see the bears, waterfalls, and titular misty peaks, the park has a lot of historical value, as well. Unlike many western parks, Great Smoky Mountains National Park was formed on what were largely privately owned lands, displacing numerous pioneer communities in the process. Most of what they left behind was cleared away or reclaimed by nature, but often the cemeteries remain as a reminder of what came before. Reading about the local history and lore of the region—things that are invaluable and under constant threat of being forgotten—I was inspired to make this map. Every name on it is an invitation to explore the story of a forgotten community, an isolated family, or a mysterious individual.

The map depicts more than 150 cemeteries found within the park's boundaries. The isometric perspective, or parallel view, prevents the sites furthest from the reader from jumbling together or getting obscured by tall peaks. Chunks of text scattered around the map help share a little bit of the local history.

Courtesy of Florida Department of Environmental Protection.

Child and infant mortality rates were much higher in the days when pioneers occupied the park. Families would often bury the deceased right on their own property, close to home. The Barnes Children, Tom Huskey's Child, Noland Children, H. D. Burris Child, and Lail Cemeteries represent some of those sites.

While slavery was not as prevalent in the Great Smoky Mountains as it was elsewhere in the American South, a handful of slave cemeteries can still be found within, including Enloe at Oconaluftee and O. E. Kerr near Cataloochee.

Humans have lived among the Great Smoky Mountains for thousands of years, with evidence of hunter-gatherer tribes dating back to 9,000 BCE. More recently, the Cherokee Nation thrived in large, permanent settlements both in and around the park up until the arrival of European-American pioneers in the 18th century.

Any graves and human remains left behind by these early cultures are protected by the Native American Graves Protection and Repatriation Act (NAGPRA). While many of these will remain forever lost, those found by park rangers and archaeologists are not likely to be shared with the general public out of respect for the dead and an effort to safeguard them from looting and vandalism.

Cemeteries within the park vary greatly in terms of size and structure. The smallest and most remote of them contain only a single grave, often with nothing more than an uninscribed rock tilted up on its end to mark the location. The largest, typically in churchyards, are neatly organized and contain several hundred graves.

Periwinkle, in addition to providing excellent ground cover and having attractive purple blooms, was once thought to ward off evil spirits. Originally native to Europe, patches of it found within the Great Smoky Mountains may indicate the presence of nearby lost graves.

Southern Appalachia has a unique tradition known as "Decoration Days," which usually occur at cemeteries around May and June. Descendants of the buried make pilgrimages to these often remote locations to honor the dead, cleaning debris and distributing flowers, in addition to singing, praying, story-telling, and "Dinner on the Ground." Around Fontana Lake, park service rangers help by transporting families to the North Shore via boat and along trails by four-wheeler, where possible.

Below Fontana Lake lie the remains of Proctor, North Carolina - one of the many communities displaced when the Fontana Dam was built to power aluminum production during WWII. Part of the relocation deal was that a new road would be built along the North Shore to provide former residents with access to the land and its cemeteries, but the highway was never completed, making this one of the most remote regions in the eastern United States.

COSBY

GREENBRIER

CATALOOCHEE

SUGARLANDS

OCONALUFTEE

DEEP CREEK

Gatlinburg

0   2.5   5   10 Miles

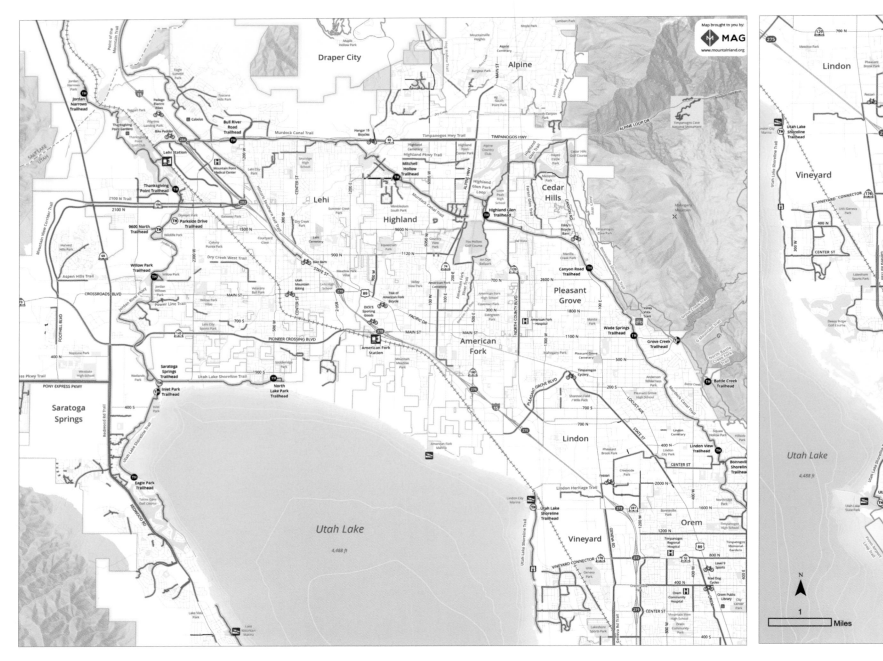

Courtesy of Mountainland Association of Governments.

UTAH VALLEY TRAILS AND BIKEWAYS

Mountainland Association
of Governments
Orem, Utah, USA
By Kory Iman

The Mountainland Association of Governments (MAG) is the regional planning agency for the Utah County metropolitan area. The association works with the public and local and state governments to provide transportation planning to meet the future needs of the region. It plays a major role in planning for and implementing a regional active transportation network that provides for nonmotorized transportation options. This map shows the existing bicycle and pedestrian system in Utah County.

MAG was challenged to develop a smaller map that could be folded down to 2.5 inches by 4 inches and resist

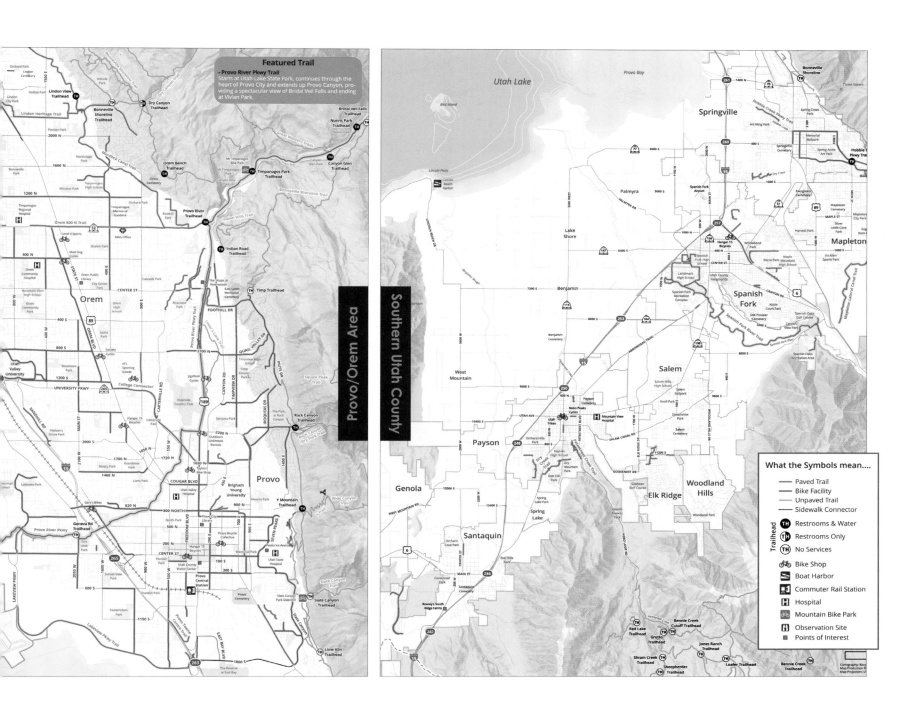

Featured Trail

- Provo River Pkwy Trail
Starts at Utah Lake State Park, continues through the heart of Provo City and extends up Provo Canyon, providing a spectacular view of Bridal Veil Falls and ending at Vivian Park.

Provo/Orem Area

Southern Utah County

What the Symbols mean....

— Paved Trail
— Bike Facility
— Unpaved Trail
— Sidewalk Connector

Trailhead
- 🅃🄷 Restrooms & Water
- 🅃🄷 Restrooms Only
- 🅃🄷 No Services
- 🚲 Bike Shop
- ⚓ Boat Harbor
- 🚆 Commuter Rail Station
- 🅷 Hospital
- 🚲 Mountain Bike Park
- 🄷 Observation Site
- ▪ Points of Interest

normal wear and tear. This was accomplished first by harnessing the power of ArcGIS Pro to develop a series of cartographic maps that clearly communicated Utah County's bicycle and pedestrian network. These maps were then paired with a pocketable layout design created using the Adobe Creative Cloud suite of tools. Last, the brochure was printed on synthetic paper to provide the durability for long-term use.

CONTACT
Kory Iman
kiman@mountainland.org

SOFTWARE

ArcGIS Pro, Adobe Illustrator,
Adobe InDesign,
Adobe Photoshop

DATA SOURCES
Utah County, AGRC, & MAG

TOPOGRAPHIC TRAIL MAP SERIES

Indiana Geological and Water Survey
Bloomington, Indiana, USA
By Matthew Johnson

CONTACT
Matthew Johnson
mrj21@indiana.edu

SOFTWARE
ArcGIS Pro, Adobe Illustrator, Adobe Photoshop, and Adobe InDesign

DATA SOURCES
2011–2013 Indiana lidar data, US Geological Survey National Hydrography Dataset (local resolution), Managed Lands in Indiana and Trails (IDNR), Street Centerlines Maintained by County Agencies in Indiana, OpenStreetMap contributors, USGS GNIS

This series of five detailed topographic maps highlights trail systems, recreational facilities, and public versus private lands along the rugged eastern edge of the Indiana Uplands region, connecting communities from Borden to Martinsville, Indiana, along the length of the Knobstone, Pioneer, and Tecumseh Trails. These maps are printed on tear-resistant waterproof paper for durability in a variety of weather conditions. Additionally, digital versions of the maps are provided through the Avenza Maps app. These printed and digital maps will support increased tourism awareness of the Indiana Upland region's recreational assets. Four of these maps were funded through a grant from Indiana University's Center for Rural Engagement.

2019 Indiana Geological and Water Survey.

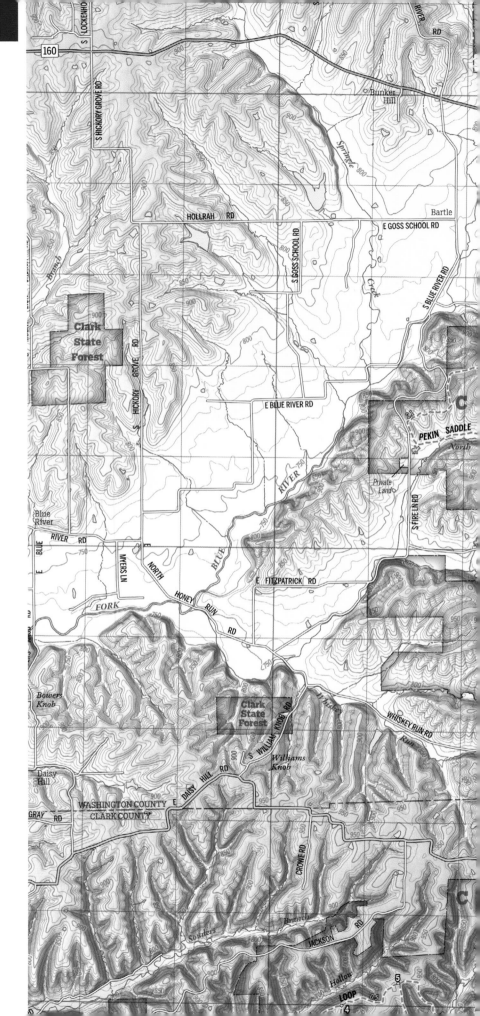

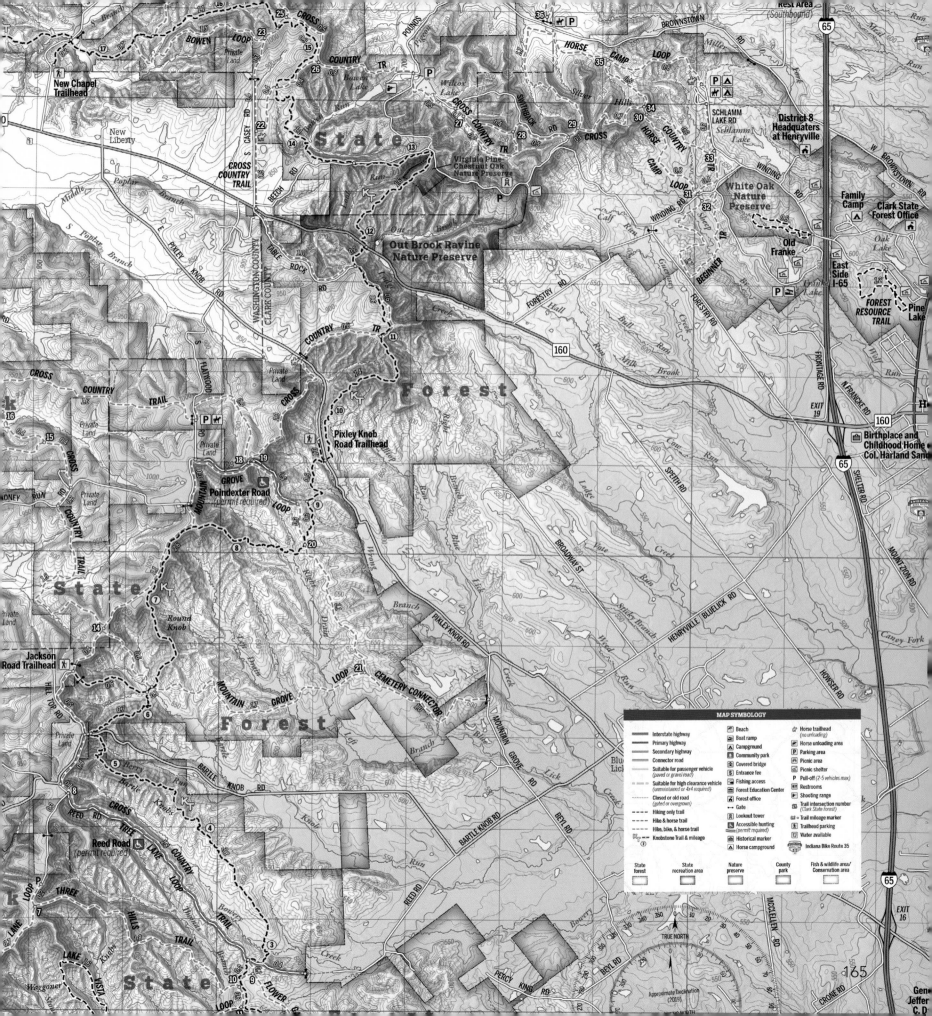

WELLINGTON: THE 10-MINUTE CITY

Wellington City Council
Wellington, New Zealand
By Órla Hammond and Marsha Badon

CONTACT
Órla Hammond
Orla.Hammond@wcc.govt.nz

SOFTWARE
ArcGIS Pro 2.4

DATA SOURCES
Wellington City Council

Wellington is home to governmental, professional, and financial services as well as New Zealand's film and special effects industries. The capital is known for its universities, museums, culture, town belt, conservation, and wind.

The city has placed at the top of Deutsche Bank's global report of the top 10 most livable cities in the world twice in a row. Wellington has quality open spaces, schools, community facilities, and public transport routes within walking distance of one another and offers opportunities to enjoy a high quality of life.

Wellington City Council wants to plan for future population growth while keeping the city compact, green, resilient, inclusive, and connected. The current population (just over 200,000) is expected to grow by 50,000 to 80,000 over the next 30 years. Wellington City Council learned that residents enjoy living close to suburban centers, where the streets are made for walking and cycling. To achieve these outcomes, the city council developed a method for implementing targeted high- and medium-density zones around suburban centers, key transportation corridors, and the central city to maintain levels of health and well-being.

This heat map shows key city amenities, including centers, public transportation routes, quality open spaces, schools, community facilities, libraries, and supermarkets. The amenities are within a five- or ten-minute walk of existing population centers, with proximity and weighting based on the anticipated distance residents could be assumed to walk.

The heat map was then used to examine the growth potential of a given area and identify the potential for further intensification around these centers. This intensification will be supported by measures to encourage a diversity of housing styles, types, and scales that satisfies the needs of future residents and contributes to the city's character and social connectedness.

Courtesy of City Design and Place Planning, Wellington City Council, New Zealand, 2020.

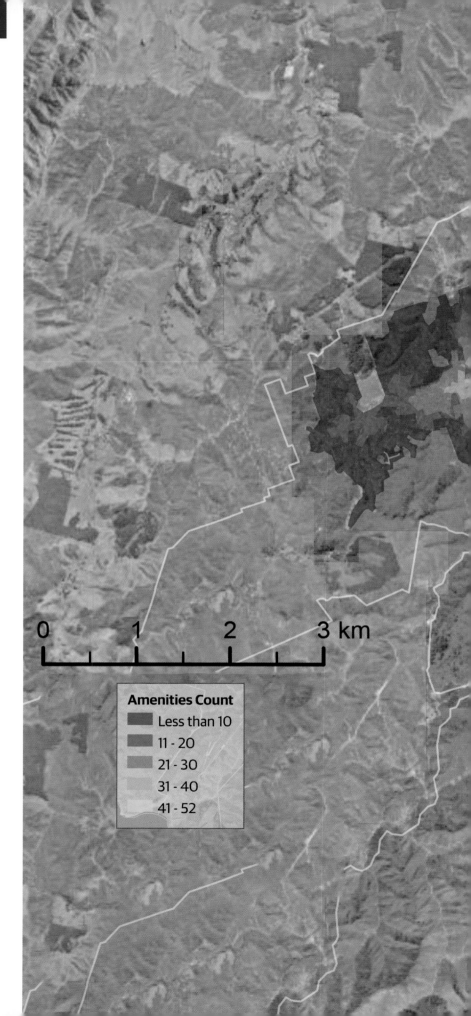

Amenities Count
Less than 10
11 - 20
21 - 30
31 - 40
41 - 52

Amenities Count

- Less than 10
- 11 - 20
- 21 - 30
- 31 - 40
- 41 - 52

0   1   2   3 km

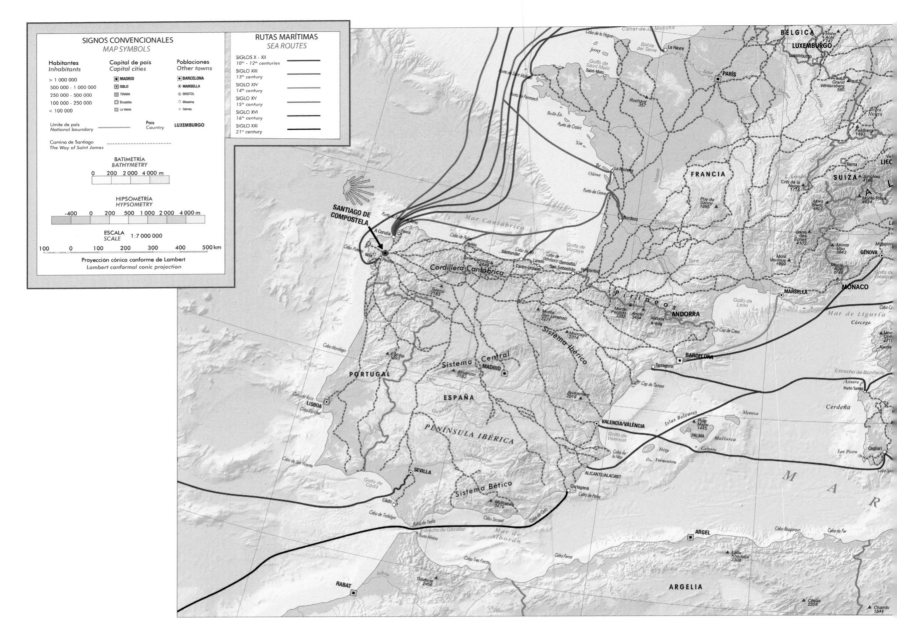

THE WAY OF SAINT JAMES ACROSS EUROPE

Instituto Geográfico Nacional
Madrid, Spain
By National Geographic
Institute of Spain

This map displays nearly 250 roads to Santiago across Europe (more than 80,000 km), contributed by 32 European Jacobean associations, and coordinated by the Spanish Federation of Associations of Friends of the Camino de Santiago. The tracks depicted on this map are marked, equipped with reception facilities for pilgrims, and documented as traditional itineraries to Santiago. Some of them are cataloged as European Cultural Itinerary or World Heritage.

The second edition of this map was created, printed, and published in 2020 by the National Geographic Institute of Spain, in cooperation with the Spanish Federation of Associations of Friends of the Camino de Santiago, Xacobeo, and Xunta de Galicia.

CONTACT
Ana Velasco Tirado
consulta@cnig.es

SOFTWARE
ArcGIS Desktop 10.7,
Adobe Creative Cloud

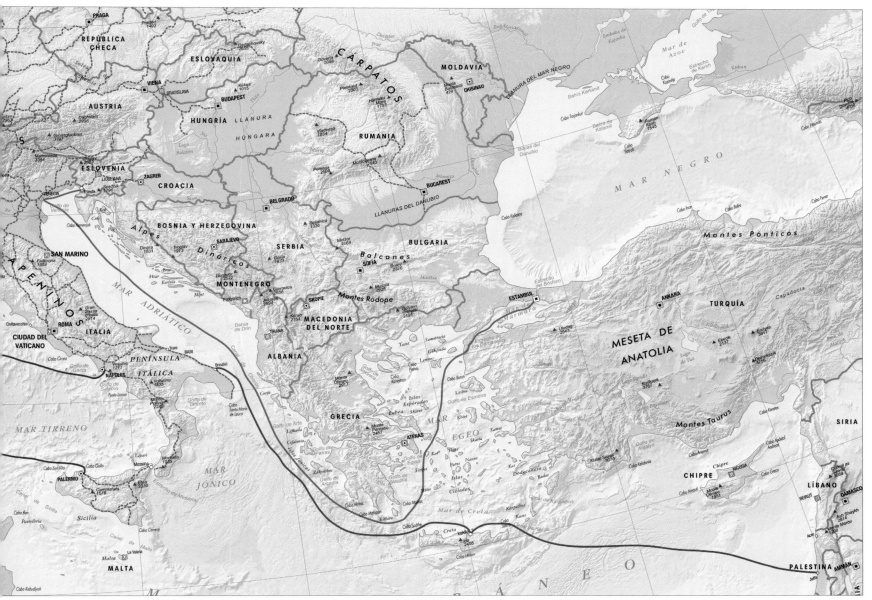

© INSTITUTO GEOGRÁFICO NACIONAL

DATA SOURCES

European Jacobean associations: **Austria**: Jakobsweg Wien; **Belgium**: Association Belge des Amis de Saint-Jacques de Compostelle Asbl, Vlaams Compostela Genootschap; **Czech Republic**: ULTREIA o.s.; **Denmark**: Danske Santiagopilgrimme; **Finland**: Jaakontien Ystävät; **France**: ACIR Compostelle, Fédération Française des Associations des Chemins de Saint Jacques de Compostelle, Szent Jakab Baráti Kör¬ (Saint James Fraternity); **Germany**: Deutsche St. Jacobus-Gesellschaft e.V., Fränkische St. Jakobus-Gesellschaft Würzburg e.V.; **Ireland**: Bernardette Cunningham, Camino Society Ireland; **Italy**: Centro Italiano di Studi Compostellani, Associazione Amici del Cammino di Santi Jacu; **Lithuania**: Šv. Jokubo Kelio Draugu Asociacija; **Luxembourg**: Frënn vum Camino de Santiago de Compostela; **Netherlands**: Nederlands Genootschap van Sint Jacob; **Poland**: Bractwo sw. Jakuba w Wieclawicach Starych, Uniwersytet M. Kopernica Wydzial Teologiczny; **Portugal**: Associação Espaço Jacobeus, Associação de peregrinos Via Lusitana, Caminhos de Santiago Alentejo e Ribatejo, http://www.santiago.com.pt, Luis do Freixo; **Romania**: Asociatia Prietenilor "Camino de Santiago"; Slovakia: Priatelia Svätojakubskej Cesty na Slovensku - Camino de Santiago; **Slovenia**: Društvo prijateljev poti sv. Jakoba v Sloveniji; **Spain**: Spanish Federation of Associations of Friends of the Camino de Santiago; **Switzerland**: Les Amis du Chemin de Sainte Jacques – Suisse;
United Kingdom: Confraternity of Saint James, Friends of the Finchale Camino.

MBTA TRANSIT DEMAND LANDSCAPE: EVENING RUSH HOUR

SPIN
San Francisco, California, USA
By Julio May

CONTACT
Julio May
julio.may@spin.pm

SOFTWARE
ArcGIS Desktop 10.7.1

DATA SOURCES
Massachusetts Bay Transportation Authority GTFS.
June 2020

The General Transit Feed Specification (GTFS) defines a common format for public transportation schedules and associated geographic information. GTFS "feeds" allow public transit agencies to publish their transit data and developers to use that data for a variety of purposes. This exercise aims to depict, on a sole image, the demand for transit services in the Roxbury–Jamaica Plain–Dorchester–Mattapan region in southern Boston, on a particular Friday evening rush hour during the summer of 2020. Spin, a company that provides dockless scooter services, uses this kind of analysis to understand the complexities of demand for transit in cities and how that demand can be complemented by e-scooters and e-bikes as an end or first-mile transportation option for citizens.

Courtesy of Spin.

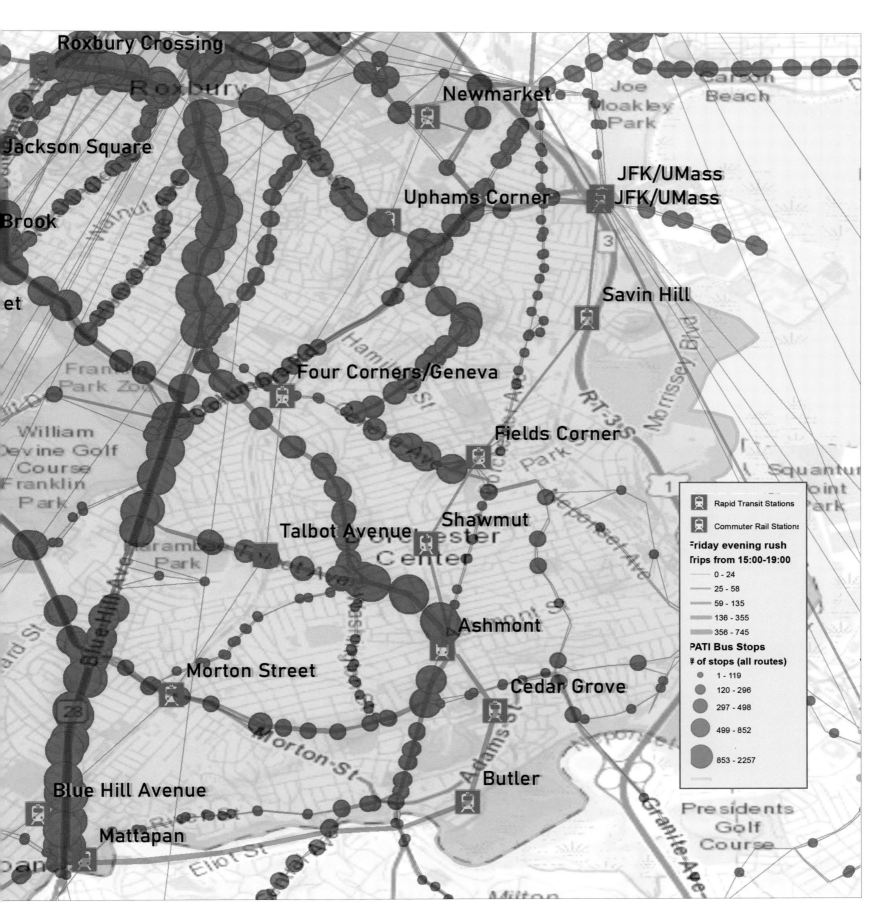

Roxbury Crossing

Jackson Square

Brook

et

Newmarket

Uphams Corner

JFK/UMass
JFK/UMass

Joe
Moakley
Park

Carson
Beach

Savin Hill

Franklin
Park Zoo

Four Corners/Geneva

William
Devine Golf
Course
Franklin
Park

Fields Corner

Shawmut

Talbot Avenue

Dorchester
Center

Ashmont

Squantum
Point
Park

Morton Street

Cedar Grove

Blue Hill Avenue

Butler

Mattapan

Presidents
Golf
Course

| | Rapid Transit Stations |
| | Commuter Rail Stations |

Friday evening rush
Trips from 15:00-19:00
— 0 - 24
— 25 - 58
— 59 - 135
— 136 - 355
— 356 - 745

PATI Bus Stops
# of stops (all routes)
· 1 - 119
· 120 - 296
○ 297 - 498
○ 499 - 852
○ 853 - 2257

ING NIGHT MARATHON 2020

Ville de Luxembourg
Luxembourg, Luxembourg
By Stefan Useldinger

CONTACT
Stefan Useldinger
stuseldinger@vdl.lu

SOFTWARE
ArcGIS API for JavaScript, ArcGIS Enterprise, ArcGIS Pro

DATA SOURCES
Ville de Luxembourg, Service Topographie et
géomatique

The 4.2-km circuit of the ING Night Marathon
Luxembourg takes runners through the Luxembourg
city districts of Kirchberg, Limpertsberg, Belair, Merl,
Hollerich, and Gare, as well as the city center. On the
evening of the marathon, around 100,000 spectators
flock to the city to cheer on the 17,000 runners and
catch a glimpse of the first person to cross the finish
line. To give the runners detailed information about
the route and its topography in advance, a public 3D
tool was created for the 2020 marathon. In addition to
details about the marathon and the half marathon, this
tool displays the kilometers, supply stations, handoff
locations for the team run, and a whole host of other
content. Visitors can locate the 500 performers and
find out about the entertainment during the event.
This year's edition of the ING Night Marathon was
canceled as part of the measures introduced due to the
COVID-19 pandemic. We hope that it will be back in
2021, bigger and better than ever.

*Copyright 2020 Ville de Luxembourg, Service Topographie
et géomatique.*

PAVEMENT WIDTHS IN THE CITY OF LONDON

City of London Corporation, London, England, United Kingdom
By Halil Siddique

	Pavement Widths in the City of London	Pavement Widths (metres)	Created by: Corporate GIS Team	0 12.5 25 50 75 Metres
CITY LONDON		──── <2m ──── 2m - 3m ──── >3m	Date Created: 05 Jun 2020	© Crown copyright and database rights 2020 OS 100023243

Courtesy of City of London Corporation.

The City of London, in the heart of the wider metropolitan area, is known as "the Square Mile." It includes much of the finance and legal central business district of London, measuring only 1.18 square miles. The residential population is small—around 8,000—but there is a large pre-COVID-19 working population of 520,000.

The task was to identify the widths of pavements to facilitate social distancing during the COVID-19 pandemic. Script was created that automated the process to generate and calculate the widths of the pavements in the city. The maps and data were then used to help the City of London Corporation's Transportation Management Team to identify bottlenecks within the pavement network, make decisions

Pavement Widths
in the
City of London

Pavement Widths (metres)	
	<2m
	2m - 3m
	>3m

Created by:
Corporate GIS Team

Date Created:
05 Jun 2020

0 50 100   200   300
Metres

© Crown copyright and
database rights 2020
OS 100023243

CITY OF LONDON

regarding the allocation of carriageway
space for walking, cycling, and queuing, and
help in changing its operation (for example,
switching to a one-way system) when
lockdown measures ease.

CONTACT
Halil Siddique
halil.siddique@cityoflondon.gov.uk

SOFTWARE
ArcGIS Desktop

DATA SOURCES
Ordnance Survey
MasterMap™

AA UK ROAD ATLAS

AA Media Limited
Basingstoke, Hampshire, United Kingdom
By AA Media Limited 2020

CONTACT

Nicky Hillenbrand
nicky.hillenbrand@aamediagroup.co.uk

SOFTWARE

ArcGIS Desktop 10.7.1, ArcGIS Production Mapping,
and Adobe Illustrator

DATA SOURCES

AA Media Limited

This map is an extract from one of the AA's 2021 UK road atlases. The map represents a departure from AA Media's standard use of Esri software to one of cartographic design in its finest form. This has been achieved through the implementation of a workflow that moves from ArcMap and ArcGIS Production Mapping to Adobe Illustrator. Using ArcMap, AA Media is designing complex symbols and creating holdouts to ensure that the text and symbols on the map are clear and easy to read. Using Adobe Illustrator, the company is creating transparency (to show one object over another) and using ArcGIS Production Mapping to generate high-quality PDF files that can go directly to the printer. AA Media is now using this workflow to create clear, well-designed maps for all AA UK road atlases.

Courtesy of © AA Media Limited 2021. Contains Ordnance Survey data © Crown copyright and database right 2021 (100021153).

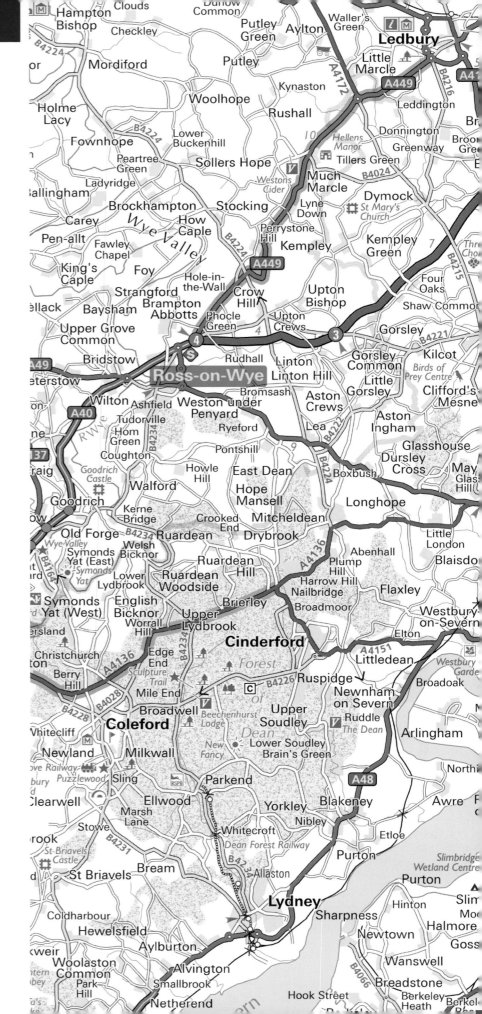

Chandlers Cross · Castlemorton Green · Norton · Ashton under Hill · Aston Somerville · Childswickham · Aston · Broad

Castlemorton Common · Longdon Heath · Uckinghall · Queenhill · Ripple · Lower Westmancote · Overbury · Grafton · Sedgeberrow · Wormington · Buckla

286 Hollybush · Birts Street · Longdon · Puckrup · Hill End · Bredon Barn · Kemerton · Conderton · Beckford · Dumbleton · Laverton · Stanton

Hollybush · Camer's Green · Rye Street · Sledge Green · Shuthonger · Bushley · Twyning Green · Bredon · Kinsham · Silk Mill · A46 · Great Washbourne · Alderton · Snowshill Manor

Kings Green · White End · Berrow · Slades Green · Bushley Green · The Mythe · Bredon's Hardwick · Aston-on-Carrant · Aston Cross · Little Washbourne · Snowshill

M50 · A438 · Tewkesbury · Walton Cardiff · Pamington · Teddington · Alstone · Toddington · Didbrook · Gloucestershire Warwickshire Railway

Pendock · Eldersfield · Linkend · Forthampton · Chaceley · Priors Park · Fiddington · Oxenton · Stanley Pontlarge · Greet · Hailes Abbey · Hailes · Farmcote

Lowbands · Lime Street · Corse Lawn · Deerhurst · Tredington · Stoke Orchard · Woolstone · Dixton · Gotherington · Langley Hill · Winchcombe

Staunton · Snig's End · Tirley · Apperley · Deerhurst Walton · Hardwicke · Bishop's Cleeve · Nottingham Hill · Woodmancote · Langley · Kineton

Brand Green · Corse · The Haw · Lower Apperley · Coombe Hill · Knightsbridge · Piff's Elm · Elmstone Hardwicke · Brockhampton · Cleeve Hill · Southam · Belas Knap

Upleadon · Hasfield · Leigh · Boddington · Uckington · Swindon · Cheltenham · Cotswold Way · Charlton Abbots · Guiting Power

Blackwellsend Green · Ashleworth · Priors Norton · Barrow · Arle · Prestbury · Hawling

Kent's Green · Hartpury · Ashleworth Quay · Bishop's Norton · Norton · Staverton · CHELTENHAM · Brockhampton

Highleadon · Sandhurst · Twigworth · Down Hatherley · Staverton Bridge · Battledown · Ham · Whittington · Sevenhampton · Syreford · Salperton

Rudford · Maisemore · Longford · Innsworth · Gloucestershire · Upper Hatherley · Charlton Kings · A40 · Andoversford · Shipton · Hazlet

Bulley · Over · Longlevens · Churchdown · The Reddings · Chargrove · Leckhampton · Dowdeswell · Foxcote · Compton Abdale · Hampnett

A40 · GLOUCESTER · Badgeworth · Leckhampton Hill · A436 · Seven Springs · Withington · Chedworth Roman Villa

Churcham · Hempsted · Barnwood · Hucclecote · Brockworth · Shurdington · Crickley Hill · Ullenwood · Coberley · Compton Abdale

Oakle Street · Minsterworth · Coney Hill · Matson · Henley · Bentham · Seven Springs · Upper Coberley · Colesbourne · Chedworth

Elmore Back · Elmore · Robinswood Hill · Upton St Leonards · Green Street · Little Witcombe · Cowley · Yanworth · Fossebridge

Farleys End · Tuffley · Prinknash · Great Witcombe · Barrow Wake · Stockwell · A435 · Colesbourne

Bollow · Quedgeley · Whaddon · Roman Villa · Birdlip · Cowley · Fossebridge

Longney · Brookthorpe · Cranham · Brimpsfield · A417 · Elkstone · Rendcomb · Col

Hardwicke · Epney · Haresfield · Rococo · Sheepscombe · Whiteway · Syde · Woodmancote · Calmsden

Upper Framilode · Moreton Valence · Haresfield Beacon · Edge · Painswick · The Camp · Miserden · Duntisbourne Abbots · Caudle Green · Winstone · North Cerney · Bagendon

Whitminster · Putloe · Standish · Pitcombe · A46 · B4070 · Miserden Park · Duntisbourne Leer · Middle Duntisbourne · Perrott's Brook

Westend · Nupend · Randwick · Stroud Green · Whiteshill · Slad · Througham · Sudgrove · Edgeworth · Duntisbourne Rouse · Daglingworth · Baunton

Claypits · Alkerton · Stonehouse · Nastend · Cashe's Green · Ebley · Stroud · Bisley · Waterlane · Bournes Green · Sapperton · Stratton · A417

Eastington · Middle Street · King's Stanley · Rodborough · Selsley · Thrupp · Bussage · Oakridge Lynch · Far Oakridge · Daglingworth · Stratton · Cirencester

Cambridge · Frocester · Leonard Stanley · Middle Yard · Woodchester · Bowbridge · Eastcombe · Oakridge · Frampton Mansell · A419 · Ampney Crucis

Coaley · Nympsfield · St Chloe · Amberley · Brimscombe · Browns Hill · Chalford · Hampton Fields · Coates · Chesterton · Harnhill · Preston · Ampney St Peter

Far Green · Uley Long Barrow · Windsoredge · Box · Burleigh · Hyde · Minchinhampton · Tarlton · Siddington · Poulto Prior · A419

Ashmead · Nailsworth · Nag's · Cherington · Ball's Green · A429

UNDERSTANDING MARITIME TRENDS IN THE ARABIAN GULF WITH RF DATA

HawkEye 360
Herndon, Virginia, USA
By Ian Avilez and Chad Margotta

CONTACT
HawkEye 360
info@he360.com

SOFTWARE
ArcGIS Pro

DATA SOURCES
Radio-frequency data

This map shows a visualization of geolocated radio frequency (RF) signals in the Arabian Gulf region. VHF marine radio signals are displayed in orange and X-band marine navigation radar signals are in green. The data was collected by HawkEye 360's satellites operating in low earth orbit.

Using ArcGIS maps, HawkEye 360 is able to illustrate RF-based activity, such as the shipping lanes of larger vessels, using X-band navigation radar, and highlights the areas where smaller vessels are still active using VHF marine radio communications. While there is a strong presence of X-band signals traveling east and west near Yemen, there aren't as many traveling north and south near Somalia. This could indicate efforts to avoid areas prone to acts of piracy and give insight into shipping patterns entering and leaving the Red Sea. This map also shows that there are instances where VHF marine channels are used over land, as shown here in Ethiopia and Yemen. Although the VHF activity identified by HawkEye 360 satellites typically indicates maritime communication, in some instances similar frequencies are used for terrestrial communication. RF data offers a new source of information to add the perspective of activity to geospatial analysis.

Courtesy of HawkEye 360.

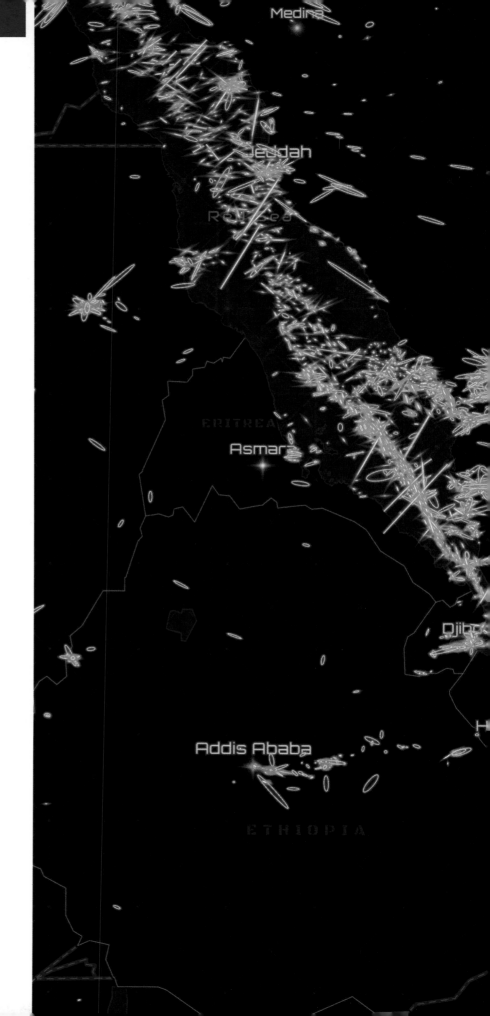

LONGVIEW TRANSIT APP AND MAPS

City of Longview
Longview, Texas, USA
By Marcus Arreguin

CONTACT
Marcus Arreguin
marreguin@longviewtexas.gov

SOFTWARE
ArcMap, ArcSDE®, ArcGIS Online, ArcGIS Collector

DATA SOURCES
City of Longview Information Services—GIS

The Longview Transit App, a product of the Longview Transit Mapping Project, shows bus routes and stops in Longview, Texas. Longview Transit serves the city with bus and paratransit services. It operates six routes that cover most of the city and run six days a week. In 2017, work began on updating our bus route and stop data, using information from Longview Transit, remote sensing, and field verification/GPS. This data was used in 2018 to create a web app that showed bus users the routes and stops. The app also helped Longview Transit with planning modifications to routes and stops.

Recently, Longview Transit made extensive changes to its routes. In response, GIS made updates to the data and the web app, at the same time enhancing the app. Users of the app can now see published bus arrival times labeled on the stops to better help them plan their commutes. In conjunction with the web app, static maps were created for download and for inclusion in the new printed map guide that Longview Transit is publishing. The static maps include a large overview map of all the routes and small single-route maps for each of the six routes. Since we made both the web app and static maps in-house, they form a unified whole by drawing on the same data and having a similar design.

The route symbology was created in ArcGIS Desktop using representation layers to help visually separate overlapping routes and show direction arrows that indicate which way buses run at a given location. Most of the field verification was completed using ArcGIS Collector. The web app was created in ArcGIS Online and uses services from our enterprise geodatabase hosted on our cloud server.

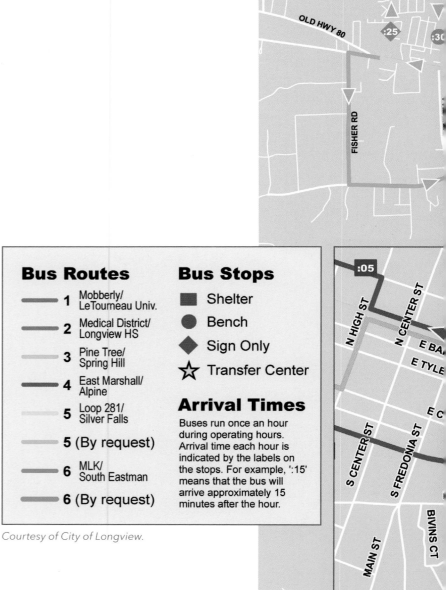

Courtesy of City of Longview.

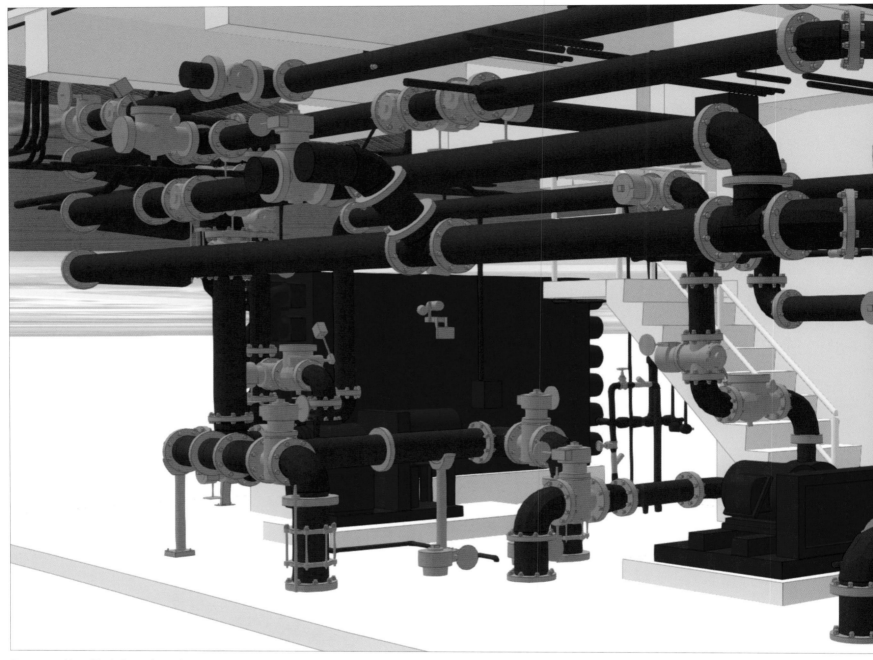

Courtesy of LandTech Consultants Inc.

3D WASTEWATER TREATMENT PLANT

LandTech
Consultants Inc.
Westford,
Massachusettes, USA
By Zachary Jaffe

Asset management is a crucial aspect of water and wastewater system management. Whether agencies are responding to an emergency leak or following an annual maintenance schedule, it's critical to know what and where the asset is. Keeping accurate asset location for indoor assets (for example, pipes and valves inside a wastewater treatment plant) is difficult. This asset information was kept on 30-year-old coffee-stained plans buried in a drawer or stored in a massive Microsoft Excel document, with no relation to the actual physical feature itself. With advancements in lidar, BIM, GIS, and asset management software, incorporating 3D data into GIS is now possible. The indoor asset data was captured in the basement of a treatment plant, via a lidar survey, modeled and brought into ArcGIS Pro, where the model was placed in its real-world location, with attributes being added to each feature.

The model was built and designed for engineers, facility/asset managers, GIS technicians, and field technicians. Office personnel use the models to identify and locate specific assets that need to be worked on and can relay that information to the field technicians. The field technicians can then use the model on-site or before going to the job site to see the asset and associated asset information.

CONTACT
Zachary Jaffe
zjaffe@landtechinc.com

SOFTWARE
AutoDesk Revit, ArcGIS Pro 2.5, VGIS

DATA SOURCES
LandTech Consultants Inc.

AURORA STORMWATER SYSTEM MAP

Aurora Water
Aurora, Colorado, USA
By Erin Courtney

CONTACT
Erin Courtney
ecourtne@auroragov.org

SOFTWARE
ArcGIS Pro 2.5

DATA SOURCES
City of Aurora, CDOT, MHFD, CDPHE, ECCV, ACWWA

Aurora Water provides safe, dependable, and sustainable water, sewer, and stormwater services to more than 386,000 residents in three Colorado counties. The storm drain system includes 387 miles of pipe, which are between 6 and 120 inches. With 48 miles of open channel, as well as 9,281 inlets and 1,867 outlets, the storm drain system is a vast and complicated network.

The Aurora Stormwater System Map provides an overview of the city of Aurora's stormwater system. Displayed in different city facilities, this wall map is used by department staff, specifically engineers and operations workers. The reference map includes basins, varied line weights, outfalls, and drop structures that provide a city scale reference to the location of critical infrastructure, as well as an understanding of the extent of the drainage system. The map highlights the tributaries, streams, rivers, and other natural and manufactured channels that transmit our stormwater safely through Aurora. An important inclusion is the floodplain data, which is helpful when planning for times of heavy precipitation.

Courtesy of Erin Courtney/Aurora Water.

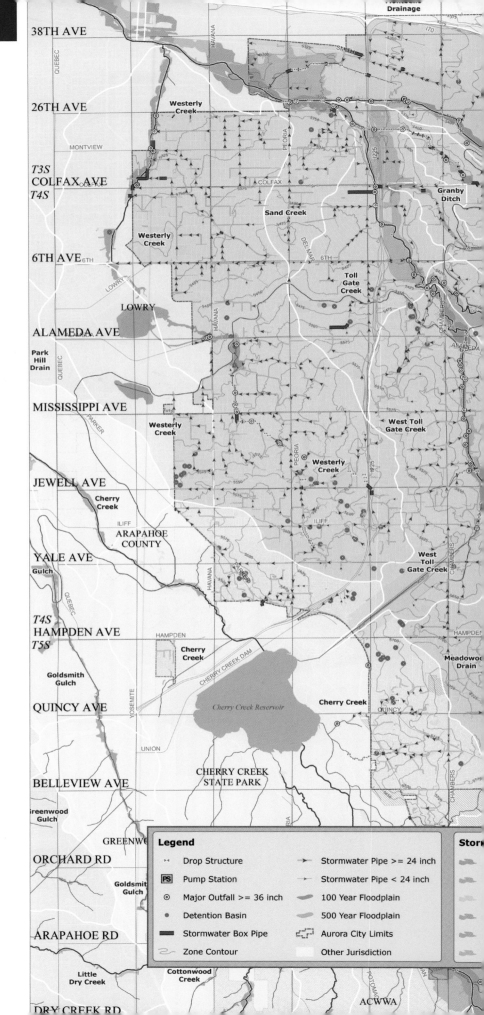

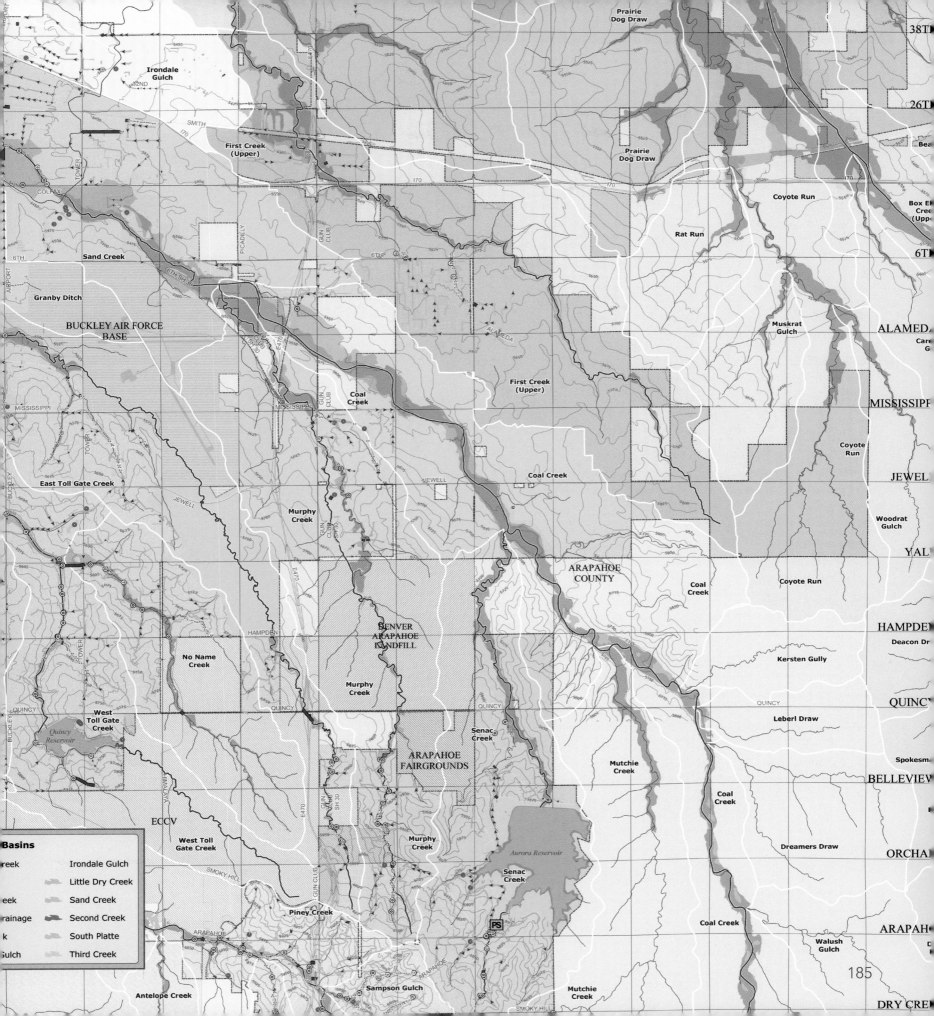

RICHLAND COUNTY INTERNET PLANNING

Revolution D Inc.
Columbia, South Carolina, USA
By James R. Stritzinger, Jr.

CONTACT

James R. Stritzinger, Jr.
jstritzinger@revolutiond.ai

SOFTWARE

ArcGIS Desktop 10.8

DATA SOURCES

Ookla®, Federal Communications Commission,
US Census Bureau

COVID-19 has laid bare the weakness of broadband infrastructure worldwide, especially in rural areas. Students are struggling to do their homework, sick patients do not have enough bandwidth to use telehealth services, and a large portion of the workforce cannot complete tasks from home.

Elected officials and community leaders are keenly aware of these challenges; however, they lack the data to make strategic investments in their infrastructure.

This project was designed to deliver tactical broadband access information to Richland County, South Carolina—home to the capital city of Columbia.

The main data layer is the result of a software model that melds the most recent Federal Communications Commission (FCC) data (internet service provider reported) together with millions of Ookla's Speedtest Intelligence® records (consumer-generated) to accurately predict the quality of internet access available in each census block.

The resultant map is fundamentally two maps in one. First, the green areas meet or exceed the FCC's recommended internet service level of 25 Mbps download / 3 Mbps upload speed, and the white areas represent regions with no homes. Finally, the remaining areas are a heat map; the darkest red or orange colors reveal the places with the highest household density and low-quality internet access.

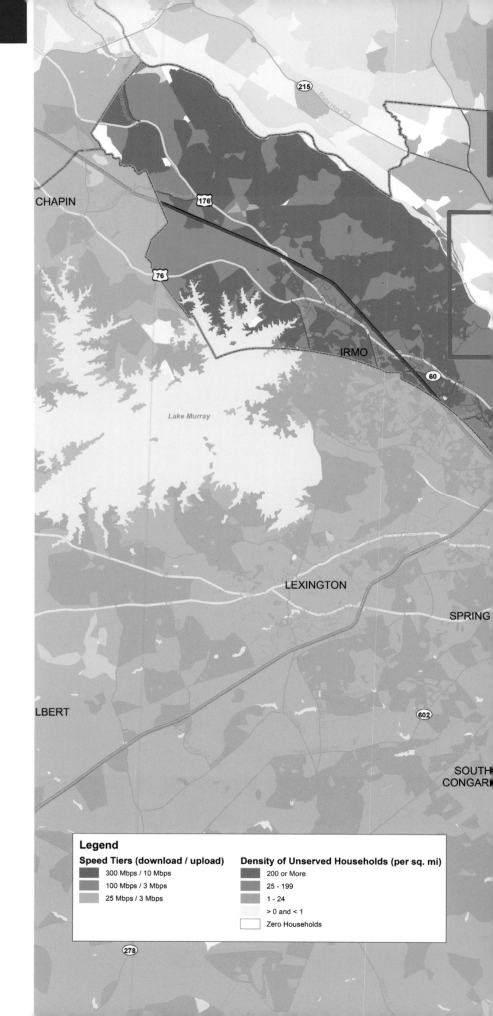

Legend

Speed Tiers (download / upload)
- 300 Mbps / 10 Mbps
- 100 Mbps / 3 Mbps
- 25 Mbps / 3 Mbps

Density of Unserved Households (per sq. mi)
- 200 or More
- 25 - 199
- 1 - 24
- > 0 and < 1
- Zero Households

LITORAL GAS DISTRIBUTION NETWORK SYSTEM

Litoral GAS S.A., Rosario, Santa Fe, Argentina, South America
By Area de Estudios y Proyectos

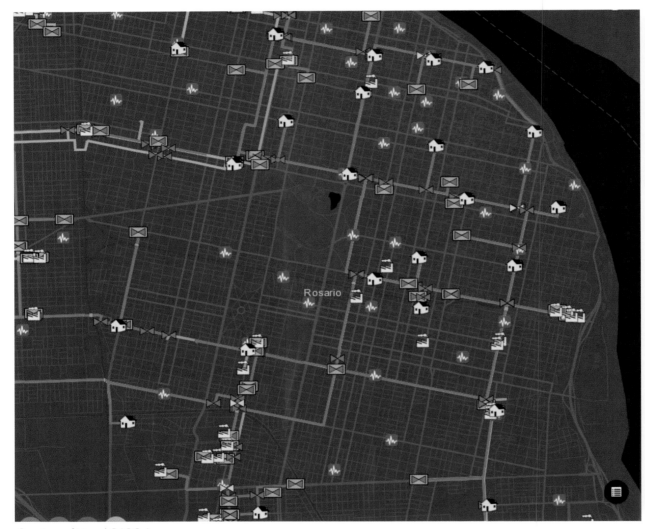

Courtesy of Litoral GAS S.A.

This web app shows the Litoral Gas distribution network system in the city of Rosario, Santa Fe Province, Argentina. It features the high-pressure gas network asset locations. It is integrated with the company's business system, MAXIMO, to grant access to pressure information and other technical data of the network assets. Fieldworkers can access the map from ArcGIS Explorer anytime, anywhere from any device in connected or disconnected environments.

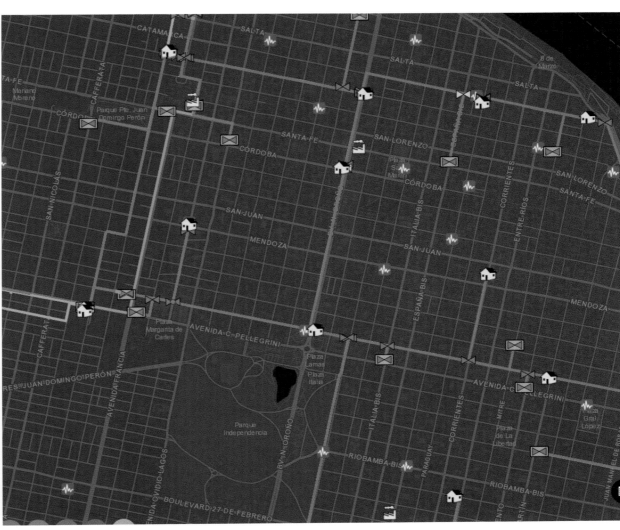

ERP

Clientes Industriales

Clientes GNC

ERP Con Propietario SDB (Instalaciones de Terceros)

ERP Con Propietario Transportista (Instalaciones de Terceros)

ESM (Instalaciones de Terceros)

Gasoductos por Presion

PRESION TRAB. (bar)
— ≤70.00
— ≤25.00
— ≤15.00

CONTACT
Silvia Bruno
silvia.bruno@litoral-gas.com.ar

SOFTWARE
ArcGIS Web AppBuilder,
ArcGIS Explorer

DATA SOURCES
Technical Management Department -
Litoral Gas S.A.

DRINKING WATER ACCESS LOCATIONS: PROVIDING AN ESSENTIAL SERVICE

Central Arkansas Water
Little Rock, Arkansas, USA
By Alex Harper

CONTACT
Alex Harper
alex.harper@carkw.com

SOFTWARE
ArcGIS Pro 2.3, ArcGIS Online, ArcGIS Web AppBuilder, Cityworks

DATA SOURCES
Central Arkansas Water (CAW), Pulaski Area Geographic Information System (PAgis)

To provide essential clean drinking water to the community, including those residential water customers whose water was disconnected because of failure to pay their bills, Central Arkansas Water (CAW) saw a need to study areas where possible free water-filling stations could be installed in the system.

The inaugural CAW-U class on leadership and supervising decided to take this research on as a class legacy project. The idea is that even though CAW is shutting off water to customers for failure to pay, CAW could and should provide safe and clean drinking water to those customers in some way in centralized locations near hot spot areas.

Using Cityworks disconnection work orders, the team was able to geolocate those locations that had been shut off the previous year. To visualize hot spot areas, ArcGIS Pro was used to generate a heat map. An ArcGIS Online web map and app were then created for everyone to view.

Along with the heat map, other information was assembled into the map, and criteria were created to determine the areas where potential water-filling stations would make the biggest impact. Current water system infrastructure and locations such as community centers, fire stations, and parks were considered. Even public transportation bus stops and parking lots were considered. Five locations were then chosen for potential water station locations.

The CAW-U Water Station Project web app provides all the data needed for upper management and commissioners to consider.

Courtesy of Central Arkansas Water (CAW), 2020.

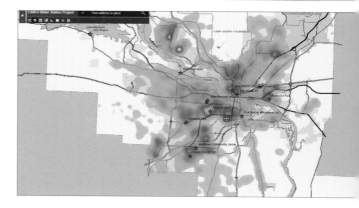

Courtesy of GeoDecisions.

NOTIFY FOR ARCGIS UTILITY NETWORK

GeoDecisions
Camp Hill, Pennsylvania, USA
By Ohan Oumoudian

We have successfully integrated our Notify solution, a high-speed mass notification system, with Esri Utility Network. Combining Notify and Utility Network allows you to identify or isolate the clients affected by a water main break or other issues identified when

using Utility Network. Notify will send text, email, or voice messages within minutes to these clients to provide additional information or directions. This integration and accuracy within a water utility network are invaluable to mobilize customers and employees

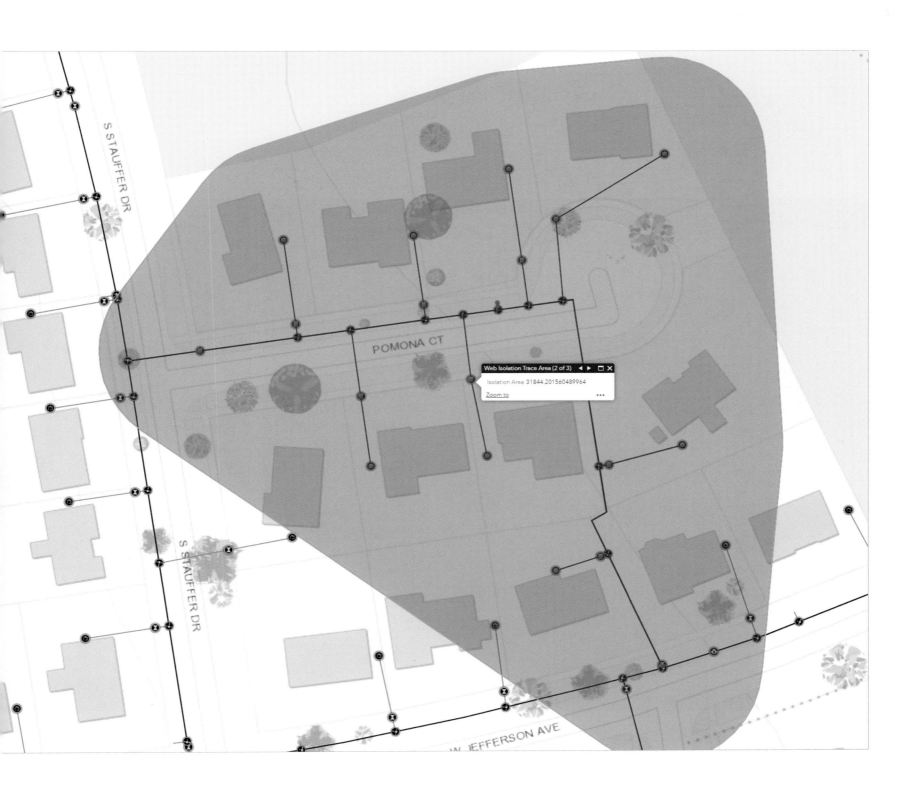

Web Isolation Trace Area (2 of 3) ◀ ▶ ▢ ✕

Isolation Area 31844.201560489964

Zoom to ···

POMONA CT

S STAUFFER DR

S STAUFFER DR

W JEFFERSON AVE

quickly and safely for any emergency. This map shows results from an isolation trace. These can be used directly by Notify for call-out campaigns from a mobile device or as authorized by the main office as part of a quality control workflow.

CONTACT
Janet Kolokithas
jkolokithas@geodecisions.com

SOFTWARE
ArcGIS for Water Utilities

DATA SOURCES
GeoDecisions

CUSTOMER MODEL DASHBOARD

Denver Water, Denver, Colorado, USA
By Jeremy Jordan, Phillip Segura, and Scott Snyders

Customer Model

Automotive	Irr - Urban Garden
Church	Master Meter
Entertainment	Mixed Use
Grocery	Multi-Family
Healthcare	Office
Hotel	Other
Ind - Brewery	Restaurant
Ind - Growhouse	Retail
Ind - Other	School - Elementary
Ind - Warehouse/Storage	School - High School
Irr - Golf Course	School - K-8
Irr - HOA	School - Middle
Irr - Other	School - University/College
Irr - Park	School Other
Irr - Public Space	Single-Family Attached
Irr - Special District	Single-Family Detached

Courtesy of Denver Water.

The Denver Water Customer Model is a combination of GIS, customer billing, and a system of characteristics that allow us to measure water use efficiency. The Customer Model boundaries were derived mostly from the official legal assessor boundary. This boundary was created by expanding the parcels to the edge of pavement to more accurately calculate square footage of irrigable areas. The building of the model was broken into manageable portions by classifying customer model types based on common characteristics necessary to measure water use efficiency. Once the Customer Model was operational and ready to be consumed, the question became how to most effectively use it. To make it more easily accessible, with little to no training, the decision was made to bring the Customer Model into the portal in the form of an operations dashboard. Using the capabilities of ArcGIS Dashboards with our Customer Model, we were able to create a powerful tool to answer questions about how water is being consumed, without the need to access multiple independent systems

and licenses. This allowed us to address various questions: how many multifamily homes are in a neighborhood? What is the consumption of a customer such as the Denver Zoo and how much area is that customer irrigating? By increasing what we know about water use, customer type, and geospatial locations, our decision-making related to our customers has been significantly improved.

CONTACT
Jeremy Jordan
jeremy.jordan@denverwater.org

SOFTWARE
ArcGIS Pro, ArcGIS Dashboards

DATA SOURCES
Data derived from various sources and compiled by Denver Water